AF878340

A WORLD OF
PREDATORS

First published in Great Britain in 2026 by NQ Publishers,
an imprint of Nextquisite Ltd.
Copyright © 2026 Nextquisite Ltd
www.nextquisite.com

All rights reserved. Unauthorised reproduction, in any manner, is prohibited.

Project Director Anne McRae | Design Marco Nardi | Text Jamie Collins
Proofreading Tayabah Khan | Layout revision Andy Mansfield
ISBN 978-1-916781-43-6

Printed in China

A WORLD OF
PREDATORS

NQ
PUBLISHERS
For enquiring minds

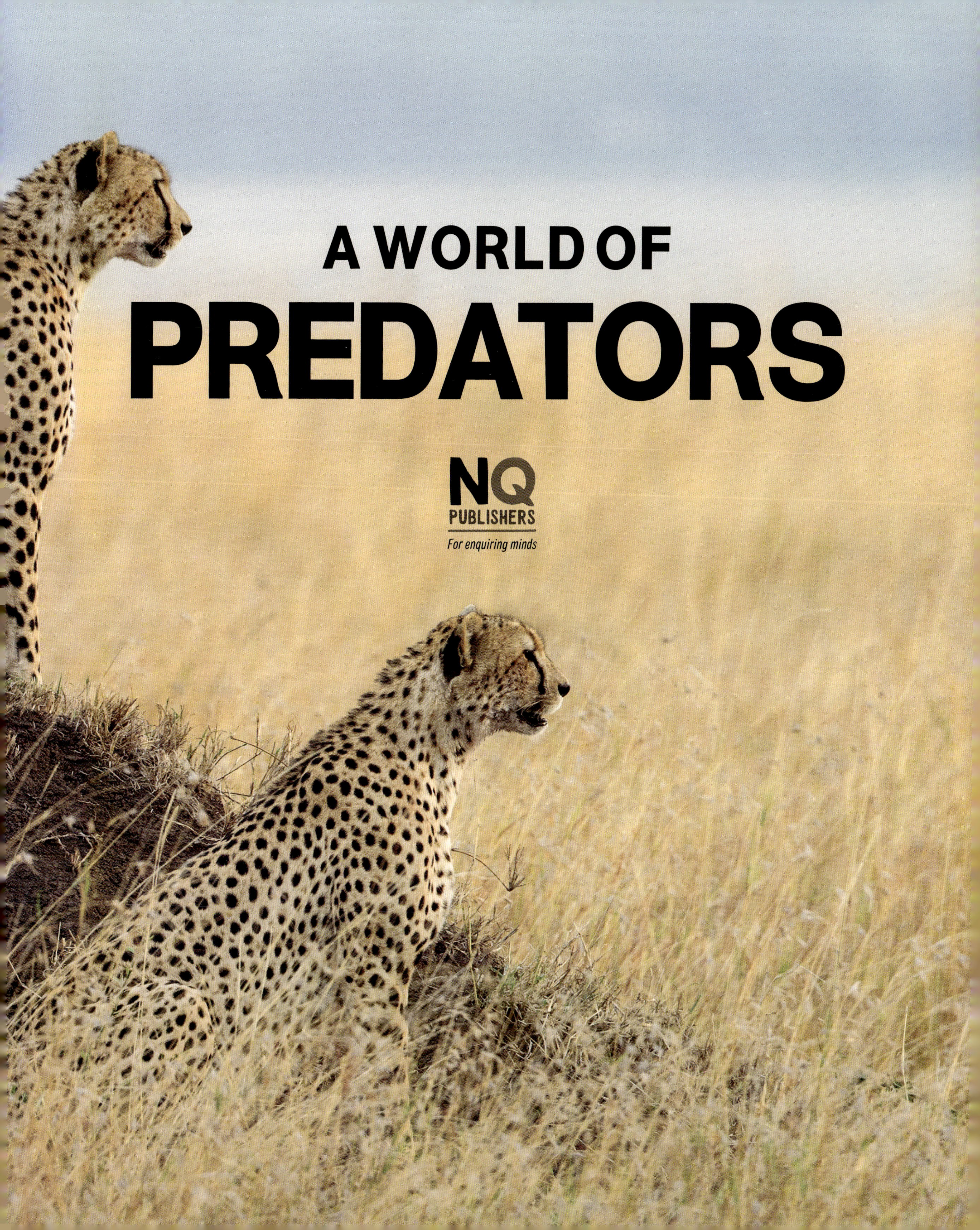

CONTENTS

NATURAL BORN KILLERS

From 'Jurassic Park' to 'Jaws', we are all fascinated by predators. A terrible chill runs down our collective spine as we watch films about apex predators like T. rex and great white sharks. In reality, nature is full of predatory animals and they play a crucial role in controlling the numbers of animals in their habitat. Their specialised hunting behaviour helps to maintain the natural balance. Today they face unprecedented threats such as climate change, habitat loss and conflict with humans over space.

The gold-ringed cat snake lives in Southeast Asia where it hunts small animals including rodents, birds, lizards, frogs and eggs. Active at night, it hunts on the forest floor after spending the day resting in the trees. It strikes using its rear fangs and venomous saliva to immobilise prey.

PREHISTORIC PREDATORS

Dinosaurs are the best-known prehistoric predators but there were many others, from huge crocodiles and sharks to giant, flightless terror birds and sabre-toothed smilodons.

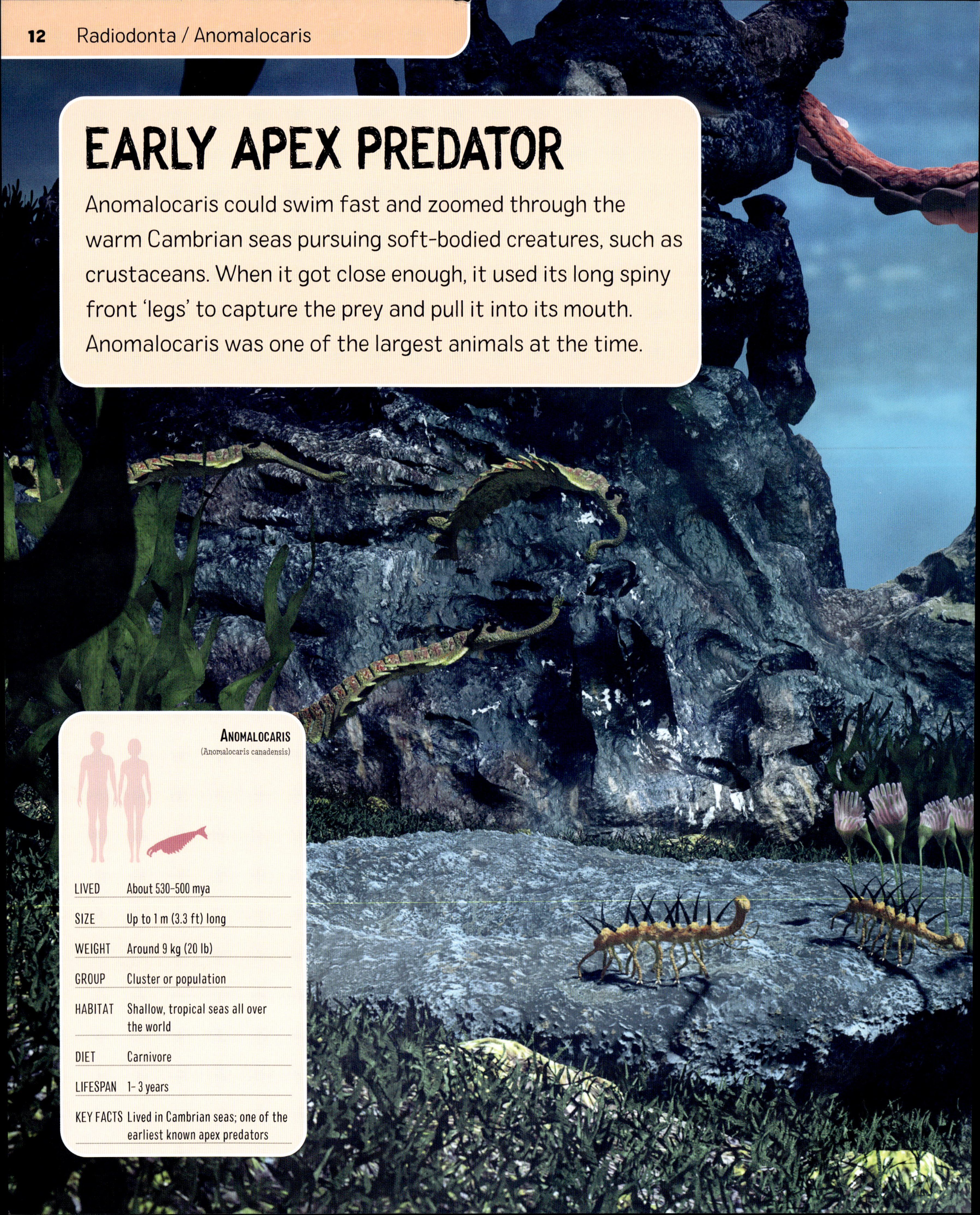

EARLY APEX PREDATOR

Anomalocaris could swim fast and zoomed through the warm Cambrian seas pursuing soft-bodied creatures, such as crustaceans. When it got close enough, it used its long spiny front 'legs' to capture the prey and pull it into its mouth. Anomalocaris was one of the largest animals at the time.

ANOMALOCARIS
(*Anomalocaris canadensis*)

LIVED	About 530–500 mya
SIZE	Up to 1 m (3.3 ft) long
WEIGHT	Around 9 kg (20 lb)
GROUP	Cluster or population
HABITAT	Shallow, tropical seas all over the world
DIET	Carnivore
LIFESPAN	1–3 years
KEY FACTS	Lived in Cambrian seas; one of the earliest known apex predators

KEEN EYESIGHT Anomalocaris had incredible compound eyes containing over 16,000 lenses perched on the end of two stalks near its head. Its vision was as sharp as modern insects, making it a formidable hunter. Anomalocaris fossils have been found all over the world, showing that it was a dominant, widespread predator.

WELL-NAMED Anomalocaris was a strange-looking creature and its scientific name actually means 'weird shrimp from Canada' in Latin! When it was first discovered, scientists thought that it probably ate trilobites, crushing their bony outer shells with its mouth parts, but they now think it ate soft-shelled creatures.

DEVONIAN SUPER PREDATOR

Dunkleosteus was a large, vicious fish covered in armour made up of interlocking bony plates. Speedy and with an extremely powerful bite, it was a top predator for about 25 million years at the end of the Devonian period.

MOVABLE NECK JOINT These huge armoured fish belong to the order Arthrodira, which means 'joined neck'. They had a joint in their necks that allowed them to raise their heads and open their mouths very wide so they could eat large prey.

MOUTH BLADES Dunkleosteus had sharp, bony blades in its mouth instead of teeth. They worked like shears, or scissors, powered by strong jaws and could slice through hard shells and heavily armoured fish.

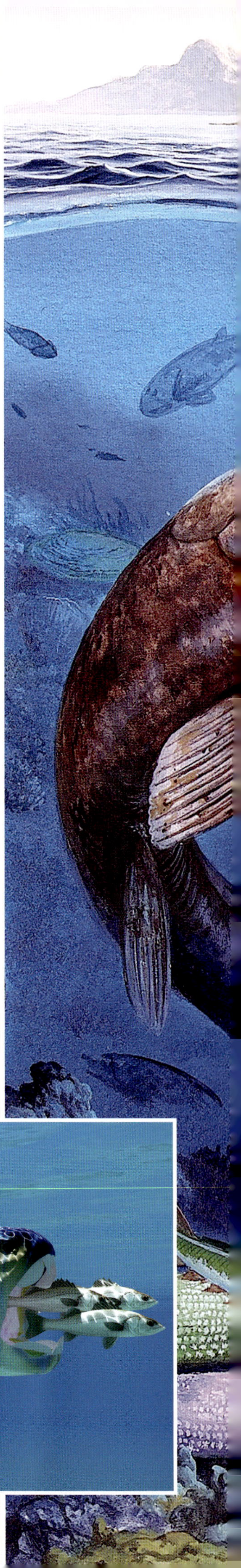

DUNKLEOSTEUS
(*Dunkleosteus terrelli*)

LIVED	About 383–358 mya
SIZE	Up 6 m (20 ft) long
WEIGHT	1–4 tonnes (1–4 tons)
GROUP	Slam of dunks
HABITAT	Shallow, warm coastal waters all over the world
DIET	Carnivore
LIFESPAN	Up to 30 years
KEY FACTS	Fast-growing; at least 10 different species; D. terrelli was the biggest; they all became extinct about 258 million years ago

BIG APPETITES Dunkleosteus fed on a wide range of prey, from sharks and other fish to ammonites (a diverse group of predatory, shelled molluscs). Fossil records also show that Dunkleosteus was a cannibal, hunting and feeding on members of its own group.

MOSASAURUS Growing up to 15 metres (50 feet) long, Mosasaurus was a short-necked pliosaur that lived in oceans around the world. They were incredibly fast, propelling themselves through the water with a powerful, shark-like tail fluke (fin). These voracious predators ate everything, from molluscs, fish and sea turtles, to sharks, birds and other plesiosaurs.

PLESIOSAURS Plesiosaurus was an early species of a large group of marine reptiles called plesiosaurs that first appeared about 203 million years ago. They were common during the Jurassic Period, but only became extinct at the end of the Cretaceous Period, about 66 million years ago.

While most plesiosaurs had very long necks, a few species – known as pliosaurs – had short necks.

ELASMOSAURUS This plesiosaur was a slow-swimming, four-flippered predator that swam in North America's shallow seas about 80 million years ago. It had one of the longest necks of any known animal, making up over half of its 12-15-metre (34-50-foot) total length. It is often associated with the Loch Ness Monster, and some people think that 'Nessie' is a surviving Elasmosaurus!

MARINE MONSTERS

Plesiosaurus lived in warm, shallow Jurassic seas. The first almost complete fossil was discovered in 1823 by British fossil-hunter Mary Anning. It was a medium size predator that mainly ate fish and squid-like creatures. It had many sharp, cone-shaped teeth in its wide jaws.

PLEISIOSAURUS

(*Plesiosaurus dolichodeirus*)

LIVED	208-183 mya
SIZE	Up to 5 m (16 ft) long
WEIGHT	Up to 100 kg (220 lb)
GROUP	Pod or school
HABITAT	Shallow coastal waters of Europe and North America
DIET	Carnivore
LIFESPAN	10-60 years
KEY FACT	Needle-sharp teeth for catching slippery prey

Scipionyx
(Scipionix samniticus)

LIVED	113–109 mya	DIET	Carnivore
SIZE	Up to 2 m (6.8 ft) long	LIFESPAN	10–30 years
WEIGHT	25–30 kg (55–65 lb)	KEY FACTS	Swift, deadly predator; small arms, big tail; big head and sharp teeth; walked on its back legs
GROUP	Flock or pack		
HABITAT	Beaches, salt marshes, lagoons of Italy		

SMALL BUT DEADLY

Scipionyx was a relatively small predator but was capable of intense bursts of speed. Its large eyes and excellent eyesight helped it to hunt fish, shellfish, lizards and insects. It lived on islands that are now part of the Mediterranean.

AN IMPORTANT FOSSIL Scipionyx was an important fossil find because it was almost whole. Not only were the bones and hard body parts preserved, but scientists could also see and examine soft body tissue such as muscles and organs. They could also see what it ate for its last meal - lizards and fish!

THE SAIL We don't really know why Spinosaurus had a two-metre (seven-foot) tall sail on its back. It may have been used to regulate its body temperature, or for displays to attract a partner at mating time.

UNDERWATER PREDATOR Spinosaurus was a strong swimmer and may even have hunted prey underwater. It had a paddle-like tail for swimming, short legs for better movement in the water, and a long snout with conical teeth perfect for catching slippery fish. Its favourite prey were giant sawfish that could grow up to eight metres (26 feet) long.

SPINOSAURUS
(Spinosaurus aegyptiacus)

LIVED	99–93.5 mya
SIZE	Up to 18 m (59 ft) long
WEIGHT	Up to 21,000 kg (46,000 lb)
GROUP	Sail, float or flock
HABITAT	Humid coastal mangrove forests, tidal flats and rivers of Africa
DIET	Carnivore
LIFESPAN	25–30 years
KEY FACTS	One of the largest carnivorous dinosaurs; huge sail on its back; spent a lot of time in the water

SAILBACK PREDATOR

Spinosaurus was a very large, semi-aquatic dinosaur that mainly hunted fish in rivers and coastal areas. It waded in shallow areas, swam along the surface in deeper waters, and may even have dived underwater to pursue prey. It had a long, narrow snout more like that of a crocodile than a dinosaur and special sensors to detect and grab fast-moving aquatic prey.

A FISHY DIET Spinosaurus was a specialist aquatic predator, hunting large fish such as sawfish and lungfish, but it also pursued aquatic reptiles, turtles, early crocodiles, pterosaurs and – on land – small dinosaurs.

GOTCHA!

The Confractosuchus lurches up from the muddy banks of the slow-moving river and snaps its big jaws firmly around the squirming body of a juvenile ornithopod dinosaur. Modern 3D scans of its stomach show that this was the prehistoric crocodile's last meal. It died even before it finished digesting it.

DIVERSITY Unlike their modern descendants (which are all dangerous predators), not all prehistoric crocodiles were apex hunters. Fossil evidence shows that some ate plants and insects, while others filtered plankton from prehistoric waters, much as many whales do today.

DEINOSUCHUS Confractosuchus was much smaller than many other prehistoric crocodiles. Here (above) we can see a huge Deinosuchus bringing down a big hadrosaur in Utah 75 million years ago. Deinosuchus grew to 11 metres (36 feet) in length and weighed up to about seven tonnes (seven tons). It was much larger than any modern crocodile.

CONFRACTOSUCHUS
(*Confractosuchus sauroktonos*)

LIVED	About 95 mya
SIZE	About 2.5 m (8.2 ft) long
WEIGHT	Up to 25 kg (55 lb)
GROUP	Bask or float
HABITAT	Queensland, Australia, semi-aquatic, warm, humid region with freshwater rivers and lakes
DIET	Carnivore
LIFESPAN	20–30 years
KEY FACTS	Famous because stomach contents show it fed on dinosaurs

A FULL STOMACH Only one Confractosuchus fossil has been discovered. It was found in Queensland, Australia, in 2010 and fully described in 2022. It was well preserved, and scientists were able to scan the contents of its stomach where they found the partially digested remains of a young dinosaur. Some experts think that the fossil belonged to a juvenile Confractosuchus and that adults may have grown much larger in size.

RAPTOR PACKS

Velociraptors were about the size of a modern wolf. They had feathers and were very fast and vicious. Due to their small size, they mainly hunted small mammals, reptiles and dinosaurs as well as insects, but they may also have worked in packs to bring down larger prey.

A VICIOUS FIGHT This illustration is based on a famous fossil find that shows a Velociraptor locked in combat with a bulky herbivorous dinosaur called Protoceratops. This proves that Velociratptor preyed on other dinosaurs. The animals were preserved in this pose because they were buried by a sandstorm or collapsing sand dune in the middle of their fight.

VELOCIRAPTOR
(Velociraptor mongoliensis)

LIVED	75–70 mya
SIZE	Up to 2 m (6.6 ft) long
WEIGHT	Up to 90 kg (200 lb)
GROUP	Pack
HABITAT	Arid, desert-like habitats in Mongolia and China
DIET	Carnivore
LIFESPAN	10–30 years
KEY FACTS	Fast, highly intelligent and feathered; had a large, sickle-shaped 'terrible claw' on each hind leg

FLYING REPTILES

Quetzalcoatlus is the largest known animal ever to fly. It was as tall as a modern giraffe and had the wingspan of a light aircraft. Although it hunted on the ground, it spent a lot of time in the air spotting new food sources from above and searching for nesting and roosting sites.

TAKE OFF Quetzalcoatlus was so heavy that scientists doubt it could lift itself off the ground to fly as a bird does. They think it probably launched from an outcrop or cliff, or vaulted a few metres into the air, flapping its wings to stay aloft.

ON THE GROUND Quetzalcoatlus mainly hunted on the ground where it stalked small animals, including reptiles, amphibians and mammals. Here we see it being pursued by two tyrannosaurids.

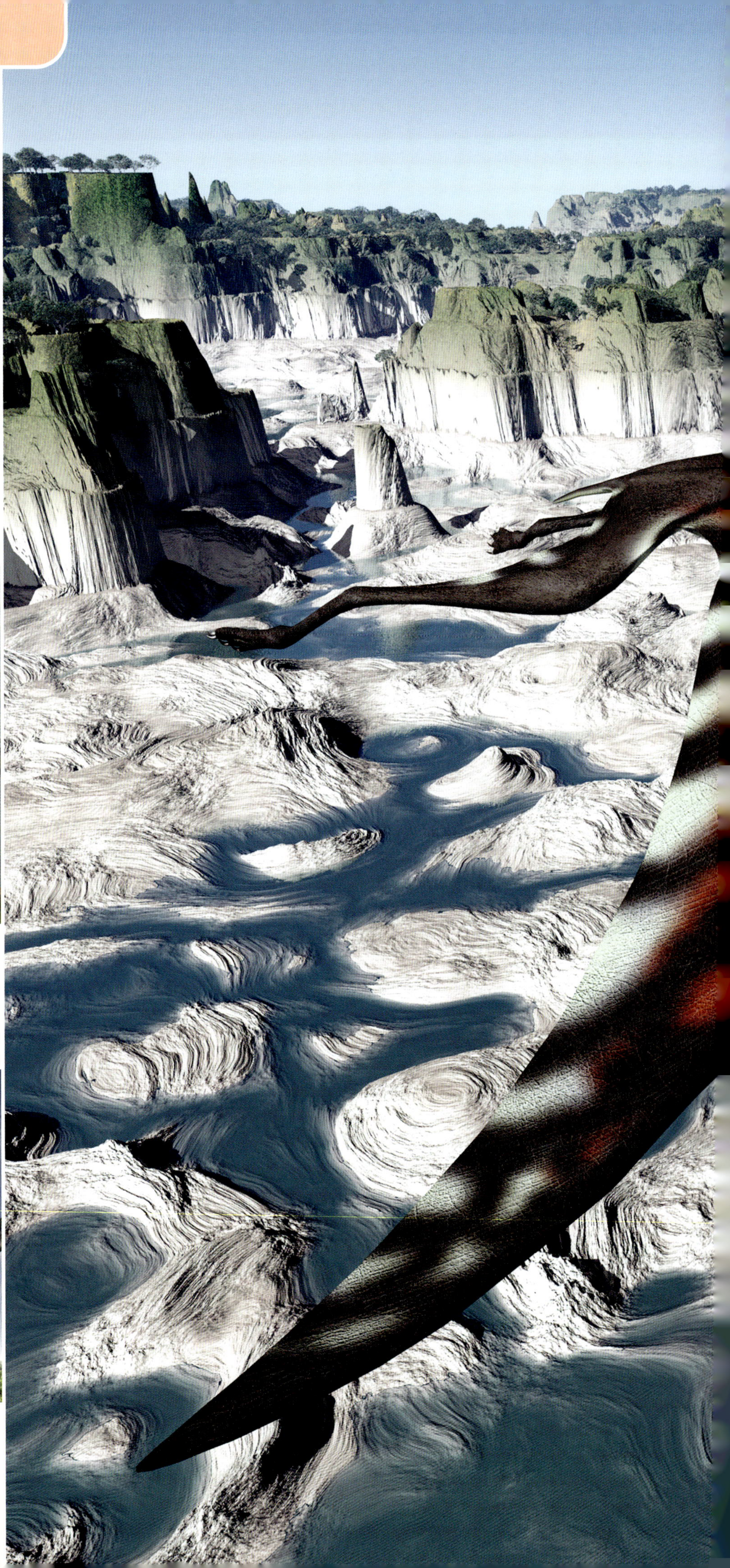

SOARING AND GLIDING Once airborne, Quetzalcoatlus was an efficient long-distance flyer and could travel hundreds of miles in a single day. Scientists think that they mainly glided and soared on air currents, rather than actively flapping their wings.

QUETZALCOATLUS
(Quetzalcoatlus northropi)

LIVED	72–66 mya	DIET	Carnivore
WEIGHT	Up to 110 kg (242 lb)	LIFESPAN	10–30 years
WINGSPAN	About 10–12 m (33–39 ft)	KEY FACTS	Largest of the pterosaurs, about the size of a light aircraft and the height of a modern giraffe
GROUP	Flock		
HABITAT	Subtropical arid plains, rivers and shallow lakes		

TYRANNIC DINOSAUR KING

Tyrannosaurus rex is probably the most famous dinosaur of all. It is widely recognised as the top predator in late Cretaceous North America, where it used stealth and ambush tactics to bring down large herbivores such as Triceratops and Ankylosaurus. Some scientists believe that it also hunted in packs to pursue massive sauropods such as Alamosaurus.

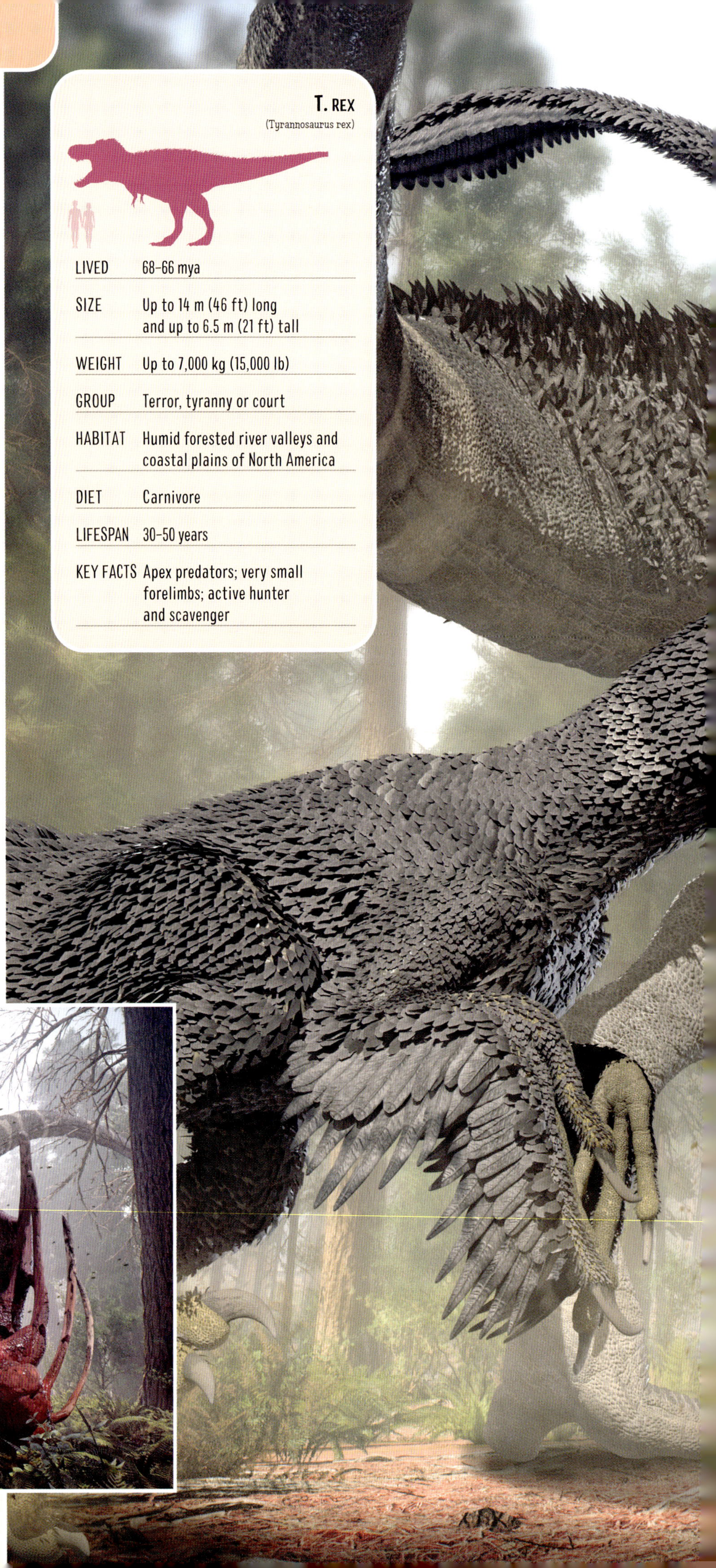

T. REX
(Tyrannosaurus rex)

LIVED	68–66 mya
SIZE	Up to 14 m (46 ft) long and up to 6.5 m (21 ft) tall
WEIGHT	Up to 7,000 kg (15,000 lb)
GROUP	Terror, tyranny or court
HABITAT	Humid forested river valleys and coastal plains of North America
DIET	Carnivore
LIFESPAN	30–50 years
KEY FACTS	Apex predators; very small forelimbs; active hunter and scavenger

HUNTER AND SCAVENGER T. rex had an excellent sense of smell and acute hearing, allowing it to detect prey that was still some distance away. Because of its immense size, it preferred to sneak up on prey, ambushing them at the last moment and using its massive bone-crushing bite power to close the deal. T. rex was also a scavenger, feasting on leftover or stolen carcasses.

DEFENCE In this frenzied scene, a fully grown T. rex is defending itself from attack by a pack of smaller Dakotaraptors. Despite being hugely outnumbered, it seems that the T. rex will win out, or if not, it will inflict a lot of damage on its assailants.

SAVAGE SNAKES

Titanoboa was the largest snake that ever slithered over the Earth. It was as long as four family cars and as round as a large hula hoop. It lived in lush rainforests and preyed on crocodiles and fish. Titanoboa ambushed its prey and wrapped itself around it. Holding the prey firmly with its sharp teeth, it crushed it to death before swallowing it whole.

TITANOBOA
(Titanoboa cerrejonensis)

LIVED	60–58 mya
SIZE	Up to 15 m (50 ft) long
WEIGHT	Up to 1,200 kg (2,600 lb)
GROUP	Nest or knot
HABITAT	Rainforests of South America
DIET	Carnivore
LIFESPAN	Around 17–20 years
KEY FACTS	Largest snake that ever lived

Titanoboa had a mouthful of sharp, backward leaning teeth that were used to grasp prey and hold them still.

APEX PREDATOR Titanoboa lived after the non-avian dinosaurs died out and was the largest predator on land for about ten million years. It had a very thick skin and, despite its weight, was a strong, fast swimmer.

MASSIVE AND MYSTERIOUS Andrewsarchus is often referred to as a 'sheep in wolf's clothing' and its appearance may have been similar to a terrifyingly large sheep-like creature with a snarling wolf's face!

BIG HEADED

Andrewsarchus was the largest carnivorous land mammal that ever lived. It roamed Mongolia about 40 million years ago. It had an odd combination of features, including hooves and a massive skull about one metre (3.3 feet) long. It jaws were packed with teeth that showed it ate meat and crushed bones. It hunted and scavenged and may also have eaten plants.

ANDREWSARCHUS
(*Andrewsarchus mongoliensis*)

LIVED	48–35 mya
SIZE	Up to 4 m (13 ft) long and 1.90 m (6.2 ft) tall
WEIGHT	Up to 1,000 kg (2,200 lb)
GROUP	Pack or herd
HABITAT	Subtropical or temperate coastal plains, river valleys and marshes of Central Asia
DIET	Carnivore
LIFESPAN	18–20 years
KEY FACTS	Strange appearance; massive size; odd – and very large – skull

WOLF-LIKE SNOUT Andrewsarchus is closely related to hippos and is a distant cousin of modern whales. Its snout was wolf-like, and its jaws were lined with large, sharp, triangular teeth used for tearing flesh. It had flat back teeth that could have been used to crush bones, or to grind up carrion, turtles and shellfish.

TERROR BIRDS

Standing up to three metres (ten feet) tall, these massive birds were the dominant predators in South America for nearly 50 million years after the non-avian dinosaurs died out. Despite their size, they could run at 50 km/h (30 mph) and had huge hooked beaks which they used to crush the skulls of their prey.

FILLING THE GAP Terror birds gradually evolved into the apex predators after the non-avian dinosaurs became extinct. South America was an island continent at the time and when the fast-running dinosaurs disappeared there was an ecological gap. Terror birds faced little competition and filled the top predator niche.

MANY SPECIES There were 18 to 25 different species of terror birds, all living in South America. When a land bridge appeared between North and South America, some species moved north.

The same land bridge eventually caused the demise of the terror birds as North American predators crossed it moving south. They became competitors for food and the terror birds eventually died out, although some species may have survived until about 18,000 years ago.

PHORUSRHACID
(*Kelenken guillermoi*)

LIVED.	43 – 1.8 mya
SIZE	Up to 3 m (10 ft) tall
WEIGHT	Up to 350 kg (770 lb)
GROUP	Flock, pack or gang
HABITAT	South America, in grasslands, savannas and open woodlands
DIET	Carnivore
LIFESPAN	About 15 years
KEY FACTS	Apex predators; largest known skull of any bird; hunted small mammals, reptiles and small birds

SUPER SHARK

Weighing as much as 30 modern great white sharks, Megalodon was one of the largest and most powerful predators that ever lived. The only marine animal ever to outweigh it is the blue whale, which is more than twice its size. Megalodons preyed on large sea creatures, including whales, dolphins, seals, other sharks and sea turtles.

WELL NAMED Megalodon means 'big tooth', which is apt, as their thick serrated teeth could grow up to 18 cm (seven inches) long. These banana-size teeth were located in jaws that were wide enough to swallow two adults standing side by side.

BITE FORCE Megalodon's powerful jaws had a bite force that was at least three times more powerful than Tyrannosaurus rex. Some scientists think they may have hunted in groups, similar to modern killer whales, to take down very large prey.

MEGALODON
(*Otodus megalodon*)

LIVED	About 23–3.6 mya
SIZE	12–18 m (40–60 ft) long
WEIGHT	About 50–75 tonnes (50–75 tons)
GROUP	Malevolence or shiver
HABITAT	Warm tropical and subtropical coastal waters worldwide
DIET	Carnivore
LIFESPAN	20–40 years
KEY FACTS	Massive size; huge teeth and powerful bite force

EXTINCTION Livyatan died out towards the end of the Miocene Epoch probably because all of its favourite food sources also became extinct due to changes in ocean currents and sea temperatures.

Livyatan
(*Livyatan melvillei*)

LIVED	About 13–10 mya
SIZE	Up to 17.5 m (56 ft) long
WEIGHT	About 57 tonnes (62.8 tons)
GROUP	Pod
HABITAT	Pacific Ocean, but possibly in oceans worldwide
DIET	Carnivore
Lifespan	12–25 years
KEY FACTS	One of the largest predators that ever lived; huge teeth and a powerful bite

FEARSOME HUNTER Livyatan's teeth grew up to 36 cm (14 in) long and are the largest biting teeth of any animal ever known. The huge extinct creature is an ancestor of the modern sperm whale, but its teeth show that it was much fiercer. Hunting by echolocation (see page 152), it was an apex predator that preyed on whales, sharks, seals, dolphins and other large sea creatures.

BATTLE OF THE SEA MONSTERS

With a surprising burst of speed, the Livyatan launches itself at the Megalodon, sinking its teeth deep into its stomach and effectively immobilising the giant shark. Livyatan is larger, faster and has bigger teeth than the fearsome Megalodon and is probably the only predator that could take it down.

SMILODONS

These powerful, predatory cats lived in the Americas up until about 10,000 years ago. They were a genus of three species with long upper canine teeth, or 'sabres'. Smilodons were heavier than today's big cats, such as lions, and were not very fast so they used ambush tactics to creep up on and surprise their prey.

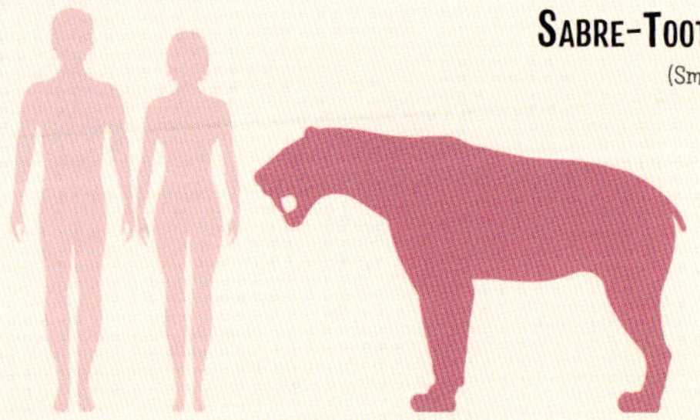

Sabre-Toothed Cats
(Smilodon fatalis)

LIVED	2.5 mya–10,000 years ago
SIZE	About 1 m (3.3 ft) tall at shoulder height
WEIGHT	Up to 280 kg (620 lb)
GROUP	Pride or pack
HABITAT	Woodlands, brushy plains and coastal areas of North America
DIET	Carnivore
LIFESPAN	10–20 years
KEY FACTS	Long, upper canine teeth, or 'sabres'; ambush hunters; lived in groups

THE BIGGEST SMILODON The largest species is known as Smilodon populator (left). It lived in South America where it hunted large herbivores, such as giant ground sloths, camels, horses and armoured caiman, rather than small, fast animals. It had very strong forelegs which it used to pin prey down before killing it with a strategic bite.

NORTH AMERICAN SMILODONS Most of what we know about these cats comes from thousands of fossils found at La Brea Tar Pits in Los Angeles, although smilodons lived all across North America. Fossils show that they were about the size of a modern lion, but sturdier and with shorter tails. They lived in social groups and probably hunted together as a pack.

In the foreground you can see two smilodon fatalis, the second-largest species. They have hunted and killed a tapir. Behind them, we can see a pack of dire wolves hunting bison.

FEARSOME DOGS

Dire wolves were about 25 per cent taller and heavier than modern grey wolves. Highly specialised carnivores, they hunted together in packs, combining cooperative stalking strategies, powerful bite force and endurance to bring down large prey, such as horses, bison, ground sloths and young mastodons. They died out at the end of the last Ice Age when large prey like these became scarce.

ANCIENT WOLVES These dire wolves have brought down a bison (right). They are defending it from grey wolves who would like to scavenge a free meal. Although the two species lived alongside one another, they were not closely related and did not interbreed.

Dire Wolf
(Aenocyon dirus)

LIVED	125,000– 10,000 years ago	HABITAT	Grasslands and plains of North and South America
SIZE	Up to 96 cm (38 in) tall at the shoulder and 150 cm (5 ft) long	DIET	Carnivore
		LIFESPAN	1–10 years
WEIGHT	Up to 110 kg (242 lb)	KEY FACTS	Pack hunters; social animals; not closely related to modern wolves
GROUP	Pack		

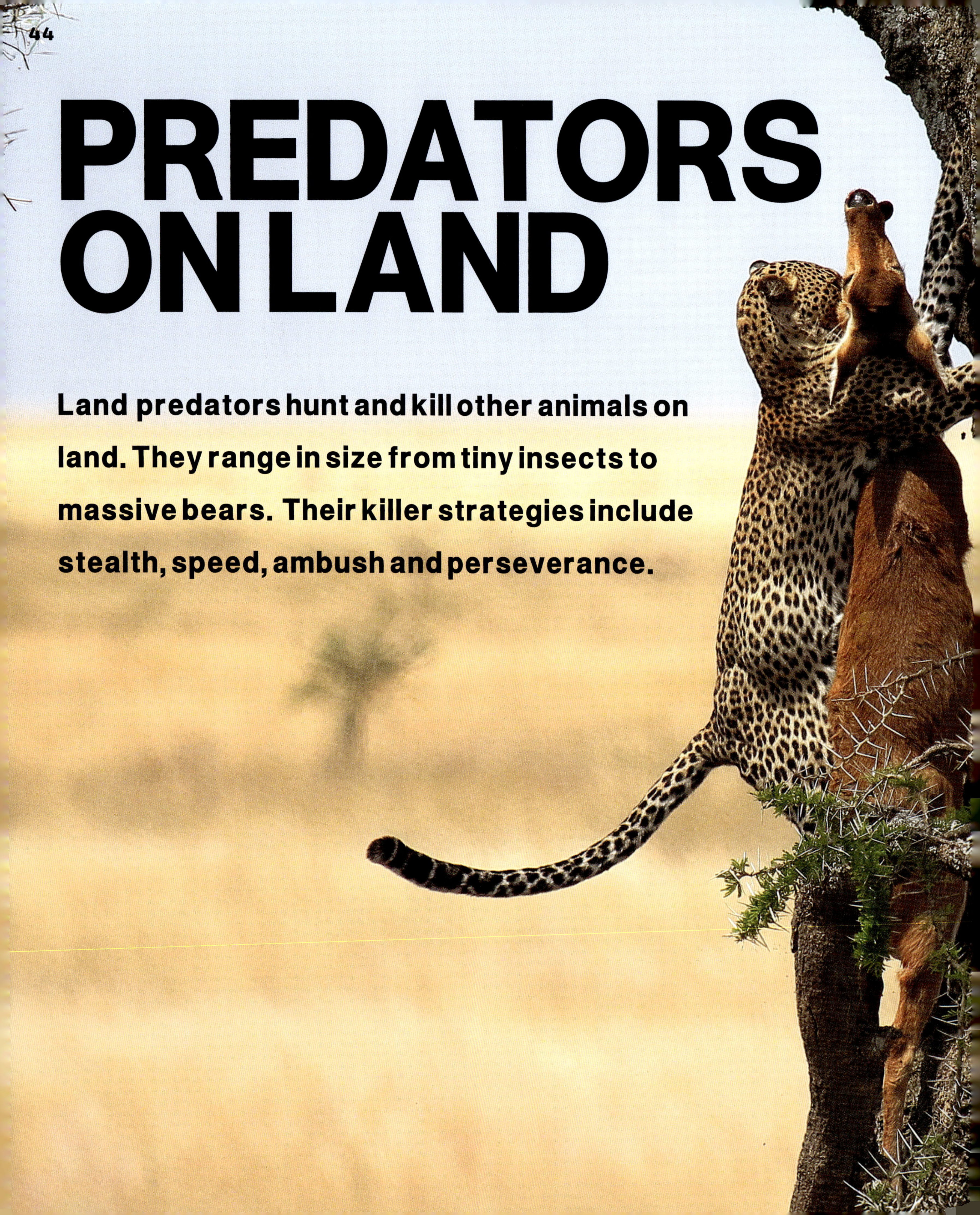

PREDATORS ON LAND

Land predators hunt and kill other animals on land. They range in size from tiny insects to massive bears. Their killer strategies include stealth, speed, ambush and perseverance.

LIGHTNING SPEED

Preying mantises are fierce predators, snatching insects from the air and catching spiders, small reptiles, amphibians and even hummingbirds with lightning speed. This one is chomping on the head of a dragonfly, using its powerful mandibles (jaws) to crunch through the insect's exoskeleton.

NOT SO APTLY NAMED Mantises are named for the way they hold their front legs in a prayer-like position, although there is no matching mercy in their hunting style! Female mantises often bite their mate's head off after – or even during – mating and nymphs (baby mantises) sometimes eat their brothers and sisters. Mantises have five eyes, but they only have one ear, which is located on their belly and can detect ultrasonic frequencies.

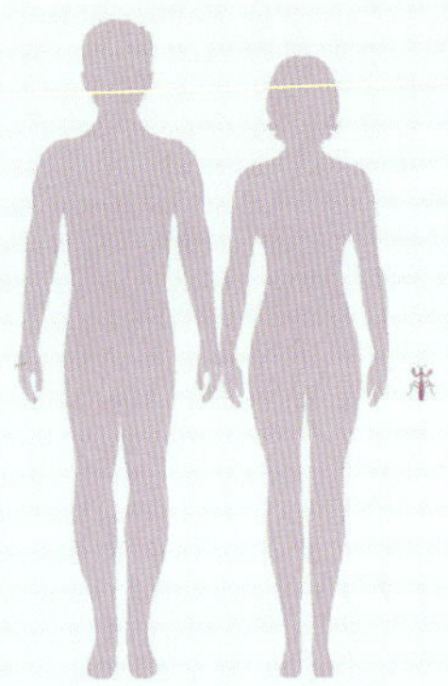

SIZE	Up to 9 cm (3.5 in) long
GROUP	Congregation
HABITAT	Grassy areas with plenty of vegetation
DIET	Carnivore
LIFESPAN	About 1 year
CON. STAT.	Least concern (globally) but threatened in some parts of Europe by habitat loss and urbanisation
KEY FACTS	Native to Europe and Asia but introduced into North America; have 5 eyes and see in 3D; only one ear; are cannibals, eating their own kind

Praying Mantis
(*Mantis religiosa*)

There are over 2400 different species of mantises. The one shown here is a European Praying mantis.

STEREOSCOPIC VISION Praying mantises have two large compound eyes made up of hundreds of tiny ommatidia (eye units). Each ommatidia has a lens, creating a wide-angle view which is perfect for detecting fast movement and gives them nearly 360° vision. They also have three smaller ocelli (simple eyes) to help them to steer while flying.

Mantises have a jointed neck which allows them to turn their heads 180 degrees. This unique feature is crucial for their hunting skills and for avoiding predators.

DANGEROUS FORELEGS Mantises have long, jointed front legs with rows of sharp spikes in strategic places so they can clamp onto prey, holding it still as they eat it live.

MASTERS OF MIMICRY Mantises use colour and posture as camouflage, mimicking leaves, flowers, twigs and bark. Some can even change colour after moulting to blend in better with new environments and ambush unsuspecting prey.

USEFUL PREDATORS

Ladybirds may look harmless and pretty but they are voracious predators. They crawl along the leaves and stems of plants using their exceptional sense of smell to locate prey, such as aphids and mites, then use their strong mandibles (jaws) to chomp them up, eating 60–80 a day. Ladybirds help keep habitats free of pests. Gardeners love them!

LIFECYCLE Like all beetles, ladybirds go through four life stages. They start out as eggs, that hatch into larvae (above). After several weeks of growing and moulting (shedding their skins), they turn into pupae (cocoons) for a few days, before emerging as adults.

MANY SPECIES There are about 5,000 species of ladybirds in the world. They vary a lot in size, colour and spot patterns. The Harlequin ladybird is larger than most and comes in a wide range of colours, from yellow to orange and red. Originally from Asia, it has been introduced to Europe and North America to control aphids and other insect pests.

Harlequin Ladybird

(*Harmonia axyridis*)

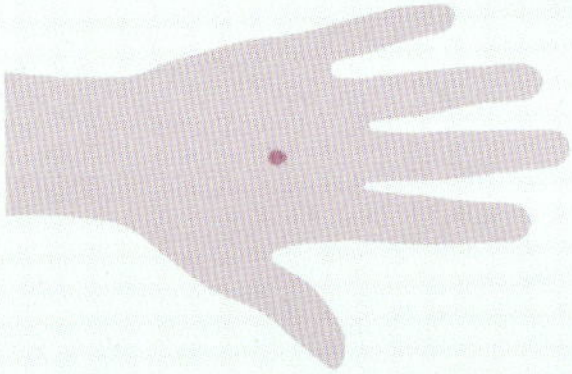

SIZE	5–8.5 mm (0.20 to 0.33 in) long
WEIGHT	0.02 g (0.0007 oz)
WINGSPAN	3–9 mm (0.11–0.35 in)
GROUP	Loveliness
HABITAT	Very adaptable, living on small shrubs, trees and grass
DIET	Carnivore
CON. STAT.	Least concern
LIFESPAN	1–2 years (complete life cycle)
KEY FACTS	Beneficial predators that keep pests in check; many different species

TOXIC TASTES Ladybirds themselves are preyed upon by a range of predators, including birds, frogs, spiders and insects. They defend themselves by releasing chemicals that taste nasty and by being brightly coloured to remind attackers that they are not nice to eat. Even so, many are eaten, especially in their younger, more vulnerable life stages.

EXPERT FLIERS Ladybirds are mainly land-based predators, however, like most insects, they have wings and can fly. They don't hunt in the air like dragonflies (see pages 154–155) but they do spend some time flying, usually to migrate, find new food sources or locate places to hibernate.

WONDERFUL MUMS Unlike insects, scorpion mothers give birth to live young rather than laying eggs. The babies are born with soft exoskeletons (outer shells) and crawl up onto their mother's backs. They stay there, under her protective stinger, for 10–20 days until their exoskeletons are hard and they are able to take care of themselves.

DEATHSTALKER SCORPION
(*Leiurus quinquestriatus*)

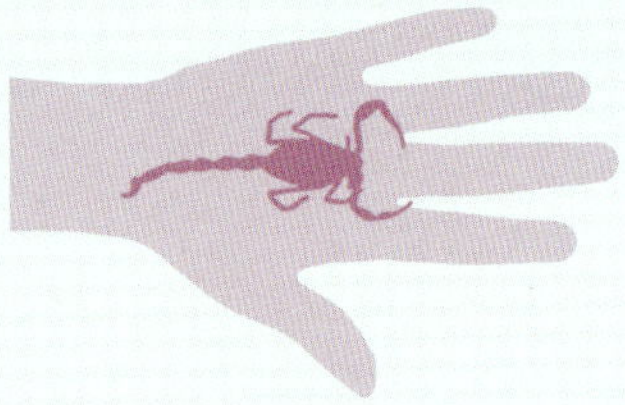

SIZE	80–110 mm (31.5–4 in) long
WEIGHT	1.0–2.5 g (0.035–0.09 oz)
GROUP	Bed, nest
HABITAT	Very dry regions of North Africa and the Middle East
DIET	Omnivore
CON. STAT.	Not evaluated
LIFESPAN	4–25 years
KEY FACTS	Uses poison to kill its prey; only a few species are deadly for humans; good mothers

NASTY STINGERS

Several thousand people die every year after being stung by scorpions. The 25 deadliest species live in North Africa, the Middle East, India, Mexico and parts of South America. In the rest of the world, a scorpion bite might make you feel ill, but won't do any permanent harm.

THE DEATHSTALKER Scorpions are ancient creatures that first appeared on Earth over 435 million years ago, long before the dinosaurs. They are all skilled predators, hunting at night and using their sharp pincers to hold prey still while they sting them with their telson (the poisonous tip of their tail). The deathstalker is the deadliest species of all. It lives in the hot, dry sands of North Africa and the Middle East. Its lethal venom is a blend of toxins. It is highly valued by the medical community who use it to treat brain tumours.

NATURE'S SQUEEZERS

These huge snakes hunt at night, ambushing prey that they locate using special heat-sensing receptors on their lips to detect warm-blooded animals. They strike with their long, backward curving fangs then coil their bodies around their victims, squeezing until they can no longer breathe. After the prey is dead, they use their highly flexible jaw to swallow the animal whole.

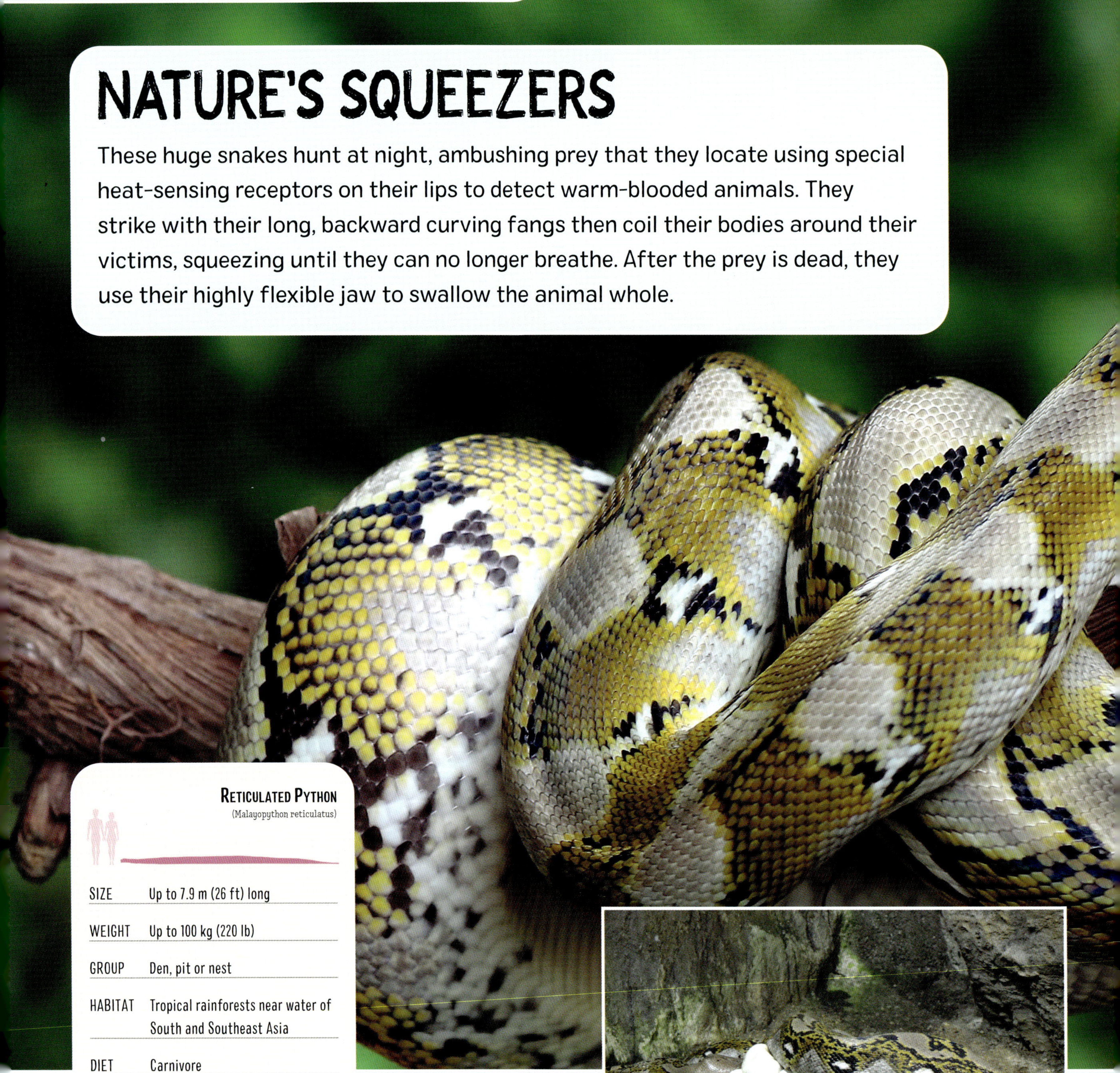

RETICULATED PYTHON
(Malayopython reticulatus)

SIZE	Up to 7.9 m (26 ft) long
WEIGHT	Up to 100 kg (220 lb)
GROUP	Den, pit or nest
HABITAT	Tropical rainforests near water of South and Southeast Asia
DIET	Carnivore
CON. STAT.	Least concern
LIFESPAN	Around 20 years in the wild
KEY FACTS	Longest snakes in the world; ambush hunters, they kill by constriction not venom

BROODING SNAKES Female pythons usually lay 15 or more eggs in a secluded spot then coil their body around them to keep them warm. It takes three months before the eggs hatch and the mother does not eat during this time.

THE LONGEST SNAKES Reticulated pythons are the longest snakes in the world. Females are longer than the males, often growing up to five metres (16 feet) long and weighing over 90 kg (198 lb). Even so, these pythons are often kept as pets. However, they are not for novice snake owners because of their size and the need for large, costly cages as well as experienced handling. Reticulated pythons are one of the few snake species in the world that do hunt and eat people.

Pet snakes are often bred in beautiful colours, like this albino individual.

NIGHTMARE LIZARDS

If Komodo dragons didn't already exist, a horror movie filmmaker would have to invent them! At up to three metres (ten feet) long – the size of a small car – these are the largest lizards in the world. They are aggressive meat-eaters, with a powerful, venomous bite. Fortunately, they only live on five remote islands in Indonesia.

TRIPOD STANCE Komodo dragons can stand up on their hind legs when they want to see further or intimidate a rival. Using their tails for support, they also stand up like this to grab out of reach prey.

WADDLING STRIDE AND FORKED TONGUES Komodo dragons walk with an ungainly, waddling stride, but they are capable of short bursts of surprising speed when pursuing prey. They flick their long, forked tongues in and out of their mouths, collecting scent particles that help them to find their prey.

KOMODO DRAGON
(*Varanus komodoensis*)

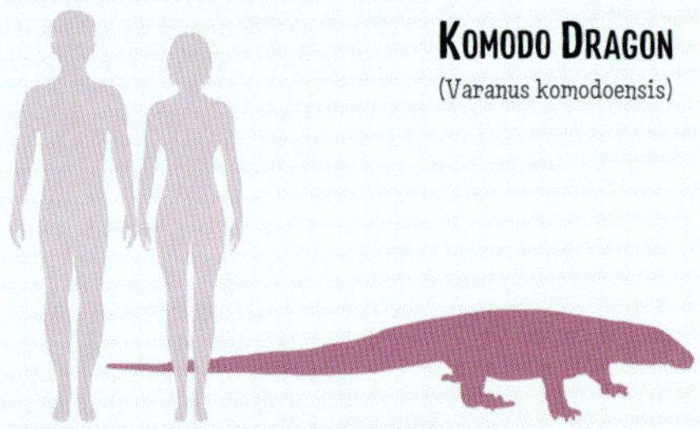

SIZE	Up to 3 m (10 ft) long
WEIGHT	70–90 kg (150–200 lb)
GROUP	Bank
HABITAT	Tropical savannas, dry monsoon forests, and coastal areas
DIET	Carnivore
CON. STAT.	Endangered
LIFESPAN	30–50 years in the wild
KEY FACTS	World's largest lizards; only lives on five islands in Indonesia; can reproduce asexually

FIGHTING DRAGONS Male dragons fight each other for dominance and access to females during the mating season. They also fight over carcasses and will sometimes eat each other, including their young. Scientists were amazed to learn that female Komodos can reproduce without male partners, in a process called parthenogenesis.

AMBUSH ARTISTS Stealthy Komodo dragons stalk their prey, ambushing them when they are close enough. They can bring down large animals such as deer, wild boars and water buffaloes and can eat 80 per cent of their bodyweight at one meal. They have sharp, serrated teeth and glands with venom that prevents their victims' blood from clotting, so that they bled out more quickly. The dragons also hunt and eat smaller lizards, snakes and birds.

DEATH ROLLERS

A massive male saltwater crocodile lurks on the banks of a brackish lagoon in northern Australia. It spends most of the day almost completely submerged, with only its eyes and nostrils showing, waiting patiently for prey. When a victim appears, the crocodile lunges forwards and grabs it in its powerful jaws then pulls it apart with a violent spinning movement known as a 'death roll'.

Saltwater Crocodile
(*Crocodylus porosus*)

SIZE	Up to 6.5 m (21 ft) long
WEIGHT	Up to 1,500 kg (3,300 lb)
GROUP	Bask, float or congregation
HABITAT	Coastal and freshwater habitats from eastern India to Southeast Asia and Northern Australia
DIET	Carnivore
CON. STAT.	Least concern
LIFESPAN	65–70 years in the wild
KEY FACTS	World's largest reptile; adapted to saltwater; massive size; strong bite force; highly intelligent

ENDLESS TEETH Saltwater crocodiles possess 64–68 cone-shaped, blunt teeth more suited to gripping and crushing prey than tearing it apart. They have one of the strongest bite forces in the animal kingdom.

BIG EATERS Saltwater crocodiles hunt at night, mainly using ambush tactics to drag prey into the water. They eat almost any animal they can overpower, from mammals, fish and birds to reptiles and crustaceans. They often hunt large prey like buffaloes, cattle, wild boars and sharks.

This saltwater crocodile is tossing a large fish into its mouth.

Nile Crocodile
(*Crocodylus niloticus*)

SIZE	3.5–5 m (81.5–16.5 ft) long
WEIGHT	Up to 750 kg (1,650 lb)
GROUP	Bask or float
HABITAT	Rivers, lakes, marshes, swamps and mangrove estuaries all over sub-Saharan Africa
DIET	Carnivore
CON. STAT.	Least concern
LIFESPAN	40–70 years in the wild
KEY FACTS	Largest predator in Africa; one of the strongest bites in the animal kingdom; highly intelligent; caring parents

TOP OF THE FOOD CHAIN Very few animals will dare to attack an adult Nile crocodile, but the youngsters are fair game for predators such as Nile monitors, mongooses and a range of different birds. Only a small number of young crocodiles survive to adulthood.

ARMOURED AND DANGEROUS

Nile crocodiles are the largest predators in Africa. With their hulking size, armour-plated skins and huge appetites, they are fearsome creatures. Every year, they attack and eat around 200 people. The males are much larger than the females, but both sexes have a very strong bite that can crush the bone and dismember the flesh of their prey in minutes.

FISHY DIET Nile crocodiles live in freshwater habitats like rivers, lakes, swamps and mangrove forests all over southern Africa. They mainly eat fish but will attack almost anything they meet, including zebras, small hippos, porcupines, birds and even other crocodiles. They are also scavengers, feeding on carrion (dead animals) and can eat up to half their body weight at a single sitting.

DEVOTED PARENTS Unlike most other reptiles, Nile crocodiles are caring, protective parents. The females lay 20–80 eggs and bury them in the sandy river banks. When the babies are ready to hatch, they make peeping sounds and the mothers dig them out. They carry the baby crocodiles in their mouths to water where they take care of them for several weeks.

ARCTIC STALKER

Male polar bears can weigh as much as ten fully grown men, so when they pounce you better watch out! One of their favourite hunting techniques is to quietly stalk a resting seal. Even though the seals are vigilant when they are out on the ice, snoozing for 30 seconds at a time then raising their heads to check for danger, the bears watch carefully, and inch forward every time the seal drops its head. Until it's too late...

Very rarely, polar bears interbreed with grizzly bears. The offspring are known as 'grolars' or 'pizzly bears'.

POLAR BEAR
(*Ursus maritimus*)

SIZE	Up to 3 m long (10 ft) long
WEIGHT	Up to 800 kg (1,750 lb)
GROUP	Pack or sleuth
HABITAT	Arctic sea ice and nearby coastal areas
DIET	Carnivore
CON. STAT.	Vulnerable
LIFESPAN	Up to 25 years in the wild
KEY FACTS	Largest bear species and largest carnivore; live mainly on ice floes in the Arctic ocean; apex hunters; very vulnerable to effects of global warming

FIERCELY PROTECTIVE MOTHERS Polar bears live alone, except for females when they have cubs. They don't breed every year, usually giving birth to one or two cubs every three years. In the autumn, pregnant bears dig dens in deep snowdrifts, where they give birth around December or January. They remain in the dens, not eating or drinking, until spring comes in April or May.

KEEN NOSES Polar bears can smell their prey up to 32 km (20 miles) away, and they can probably smell seals up to 1 km (3,200 feet) beneath the pack ice. In spring, the bears feast on seal cubs too small to escape their tactics. Later in the year, hunting gets harder and the bears scavenge other predator's leftovers or eat small birds, eggs, berries and kelp.

THE WAITING GAME Polar bears mainly eat seals and they spend about half their lives hunting them. A common technique is to sit and wait by a seal's breathing hole or at the water's edge. The bears know that sooner or later the seals have to come up to breathe, so they wait patiently, sitting still for up to 10–12 hours, until a seal pops its head above water.

THE BEAR FACTS

Brown bears are large, powerful predators. They are very intelligent and have a highly developed sense of smell that allows them to detect prey that is miles away. Brown bears hibernate for five to six months each year. Despite their huge size, they can run fast and are excellent swimmers.

HUNTERS When they hunt, brown bears go after large prey like the calves of moose, deer and caribou or smaller creatures such as ground squirrels, marmots and mice. They also consume insects such as moths.

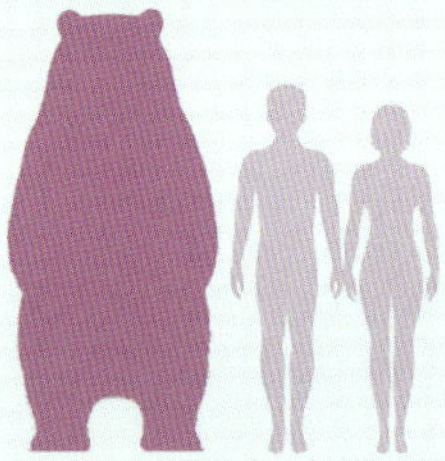

Brown Bear
(*Ursus arctos*)

SIZE	Up to 3 m (10 ft) tall (Standing upright)
WEIGHT	Up to 640 kg (1,400 lb)
GROUP	Sloth, pack, maul or sleuth
HABITAT	Forests and mountains of North America, Europe and Asia
DIET	Omnivore
CON. STAT.	Least concern
LIFESPAN	About 20–25 years in the wild
KEY FACTS	Solitary, except during the salmon run; eat a wide variety of food

FLAT-FOOTED Like humans, brown bears walk flat-footed with their heels touching the ground. They are also able to stand up on their hind legs when they want a better view or to catch a scent.

NOT PICKY EATERS Brown bears are omnivores and they eat a lot of different food. Even though they are big and strong, about 90 per cent of their diet is made up of berries, roots, nuts, fungi and insects.

THE SALMON RUN Every summer on the Pacific coasts of North America and Asia millions of salmon return to their breeding grounds to spawn. Brown bears know this and they gather along the rivers in increasing numbers from June to October to catch them. The bears often selectively eat the most nutritious parts of the fish, such as the skin, brain and eggs, rather than the whole thing. The weight they put on sees them through their long winter slumber.

PACK HUNTERS

A pack of African wild dogs surrounds a desperate warthog. The wild dogs spotted that the elderly female warthog was limping and lagging behind the others in her group and have been tracking her for hours. They have cut her off from her herd and are closing in for the kill.

FAMILY GROUPS African wild dogs, also known as Cape hunting dogs, live together in family packs led by an alpha male and female pair. Usually, only this pair breeds, producing litters of up to 20 pups, which the entire pack helps to raise. Pack members share food, often regurgitating meat for pups and the elderly, sick or injured.

African Wild Dog

(Lycaon pictus)

SIZE	72–112 cm (28–44 in) long + 29–41 cm (11–16 in) tail
WEIGHT	18–36 kg (40–80 lb)
GROUP	Pack
HABITAT	Plains, savannas, arid zones and woodlands in sub-Saharan Africa
DIET	Carnivore
CON. STAT.	Endangered
LIFESPAN	10–12 years in the wild
KEY FACTS	Live in family packs of 7–15 individuals; use vocalisations to coordinate hunting

PACK MENTALITY With a 60–90 per cent success rate, African wild dogs are very efficient hunters. They use a combination of endurance, speed and cooperation to bring down prey that is often much larger than themselves. During the hunt, they 'talk', using bird-like twittering, whines and squeaks to coordinate their attack. They eat quickly after the kill, with pups often feeding first, followed by adults. They don't want to give larger predators such as hyenas time to steal their food.

ANYTHING GOES

Foxes are skilled predators as well as scavengers. They mainly hunt small mammals, such as rabbits and voles. This one has caught a rabbit and has a satisfied expression as it heads back to its den to eat it. It will store any leftovers for the next day.

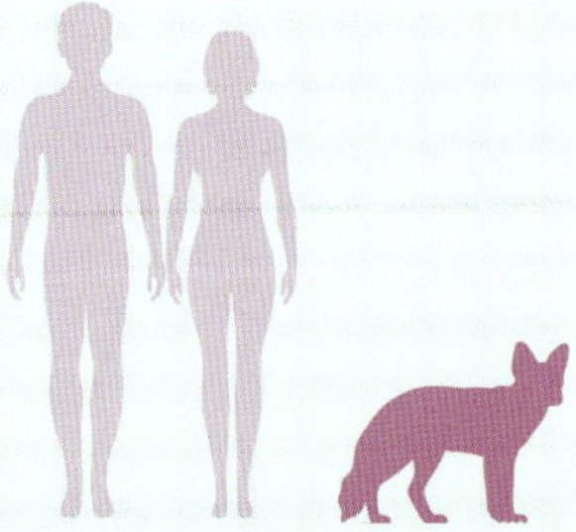

RED FOX
(Vulpes vulpes)

SIZE	Up to 86 cm (34 in) long + 55 cm (22 in) tail
WEIGHT	Up to about 12 kg (26 b)
GROUP	Skulk or leash
HABITAT	Wide variety of habitats, including farms and cities
DIET	Omnivore
CON. STAT.	Least concern
LIFESPAN	2–4 years in the wild
KEY FACTS	Largest and most common of the true foxes; skilled hunters but also scavengers; increasingly live in cities

POUNCERS Foxes have good hearing which they use to locate prey hiding under snow or grass. They leap into the air then dive headfirst, striking the prey with their nose, before grabbing it.

A VERSATILE DIET Although foxes have the teeth of a carnivore, their digestive system allows for the consumption of plant matter. As well as small mammals, birds and insects, their diet frequently includes berries, fruits, vegetables and seeds. Urban foxes feast from dustbins, bird feeders and compost heaps.

Foxes have also been observed catching fish, usually during the spawning season when large fish appear in shallow inland waters.

NOISY CREATURES Foxes are very vocal, and use more than 12 distinct sounds to communicate, including whines, squeals and barks. Like pet cats, foxes are also known to purr when they are happy.

SOCIAL ANIMALS Dingoes live together in family packs, usually made up of an alpha pair (male and female) and their offspring. They often hunt cooperatively, bringing down animals much larger than themselves, including red kangaroos, emus and even water buffaloes. They usually hunt smaller prey on their own.

STARTING YOUNG Dingoes breed every year between March and June, producing four to six pups. The youngsters are weaned at about two months, although they often stay with their parents for up to a year. At about three months, they start joining adults on hunting expeditions, learning how to chase, ambush and kill prey.

Dingoes have lived in Australia for at least 4,000 years. They are closely related to other wild dog species in Southeast Asia. Many believe that they were introduced to Australia by Asian seafarers.

TOP DOGS

Dingoes are a type of wild dog that live in Australia where they are the top land predator. They hunt a wide variety of different animals, including large mammals such as kangaroos, wallabies, wombats and possums, along with rabbits, birds, reptiles, fish and insects. Although they are mainly carnivores, they also eat some grasses and fruit.

SKINNY DOGS Dingoes are well adapted to the arid conditions in Australia. They are naturally muscular and lean and can survive for long periods without water. They help to promote biodiversity in their habitats, boosting the populations of native animals by controlling the numbers of non-native predators such as wild cats and foxes.

COASTAL DIET Dingoes that live close to the sea, like those on the Queensland island of K'gari (formerly Fraser Island), hunt and scavenge a lot of seafood, such as crabs, fish and turtle eggs. They have even been seen feasting on large dead whales.

DINGO
(*Canis lupus dingo*)

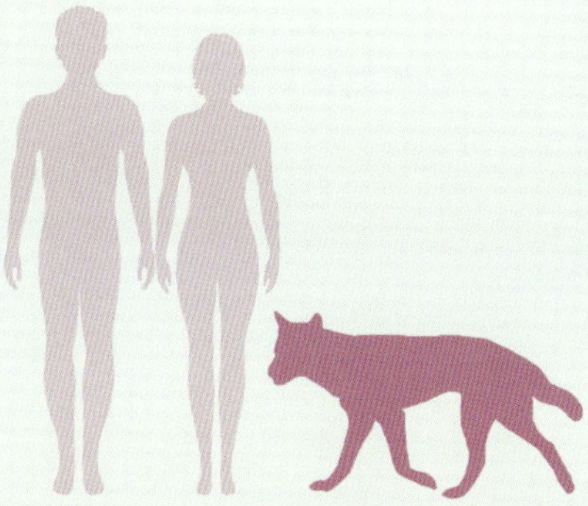

SIZE	Up to 1.6 m (5.2 ft) long, including their tail
WEIGHT	Up to 25 kg (55 lb)
GROUP	Pack
HABITAT	Deserts, forests, grasslands and uplands of Australia
DIET	Carnivore
CON. STAT.	Not evaluated
LIFESPAN	Up to 14 years in the wild
KEY FACTS	Native to Australia; probably descended from domesticated dogs in Southeast Asia that became wild again; top land predator in Australia

DOMESTIC DOG ANCESTORS
Grey wolves are the ancestors of all our domestic (pet) dog breeds. They share 98.8 per cent of their DNA and can interbreed.

THREE SPECIES
There are three species of wolves: grey wolves, red wolves and Ethiopian wolves. Grey wolves are by far the most common species and about 30 different subspecies are recognised. Here you can see two Arctic grey wolves devouring food in the snow. They are a common subspecies and live in North America and Greenland.

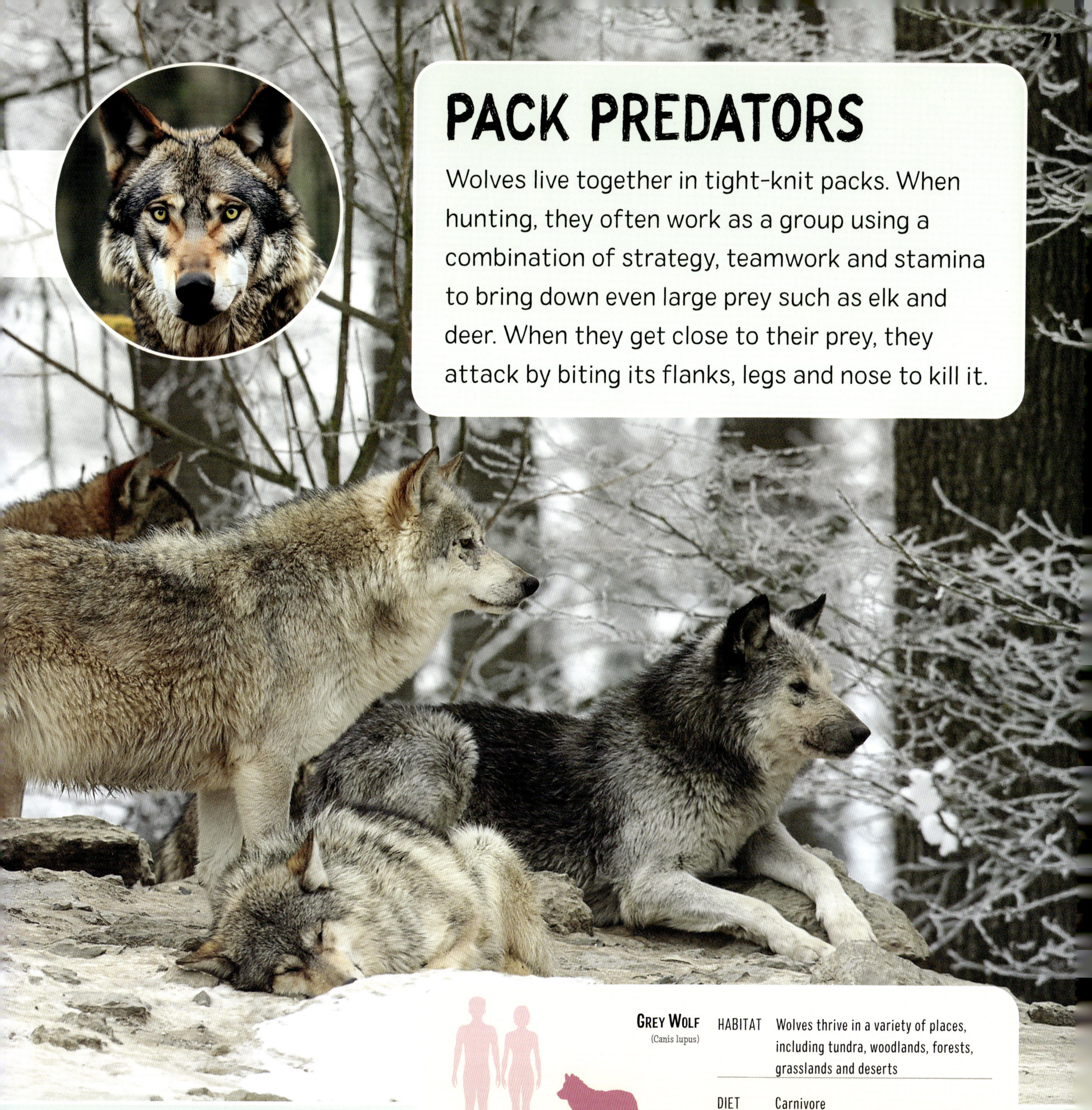

PACK PREDATORS

Wolves live together in tight-knit packs. When hunting, they often work as a group using a combination of strategy, teamwork and stamina to bring down even large prey such as elk and deer. When they get close to their prey, they attack by biting its flanks, legs and nose to kill it.

WOLFISH PROVERB

'For the strength of the Pack is the Wolf, and the strength of the Wolf is the Pack,' from Rudyard Kipling's *The Jungle Book*.

GREY WOLF
(Canis lupus)

SIZE	Up to 85 cm (33 in) tall at the shoulders and up to 160 cm (63 in) long
WEIGHT	About 40 kg (88 lb)
GROUP	Pack
HABITAT	Wolves thrive in a variety of places, including tundra, woodlands, forests, grasslands and deserts
DIET	Carnivore
LIFESPAN	Up to 13 years in the wild
CON. STAT.	Least concern
KEY FACTS	World's largest dog species; good hunters; claim large territories; live in close-knit family packs

ADAPTING TO THE SNOW Snow leopards live high in the mountains of South and Central Asia. They are well adapted to this extreme environment and have thick, warm coats and a big bushy tail which they wrap around their bodies for extra warmth. Their wide paws help them balance in the slippery conditions and are covered with fur to protect their feet from the icy ground. While other big cats have yellow eyes, snow leopards have bright blue, green or grey eyes to help them blend in with their surroundings.

GHOST CATS

This snow leopard has been tracking its prey for hours. Staying low and using its dappled grey fur to blend into its surroundings, it waits until it is close enough to pounce. When it does, the victim doesn't stand a chance! This stealth, combined with its colouring and behaviour, make it very hard to spot and have earned it the nickname of 'ghost cat', or 'ghost of the mountains'.

SNOW LEOPARDS
(*Panthera uncia*)

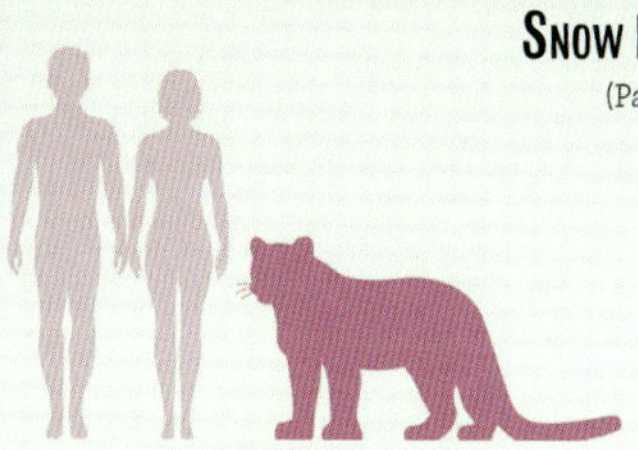

SIZE	Up to 2.1 m (7 ft) long, including the tail
WEIGHT	Up to 60 kg (132 lb)
GROUP	Not applicable; they live alone, not in groups
HABITAT	High mountains of Central and South Asia
DIET	Carnivore
CON. STAT.	Vulnerable
LIFESPAN	15–18 years in the wild
KEY FACTS	Solitary, except during the salmon run; eat a wide variety of food

PEAK PREDATORS Don't be fooled by this cub's sweet expression – it will soon grow into a powerful hunter. Snow leopards are the top carnivores of their ecosystem. They feed on blue sheep, ibex, marmots, hares, small rodents and game birds. Despite their name, snow leopards are actually more closely related to tigers than to leopards.

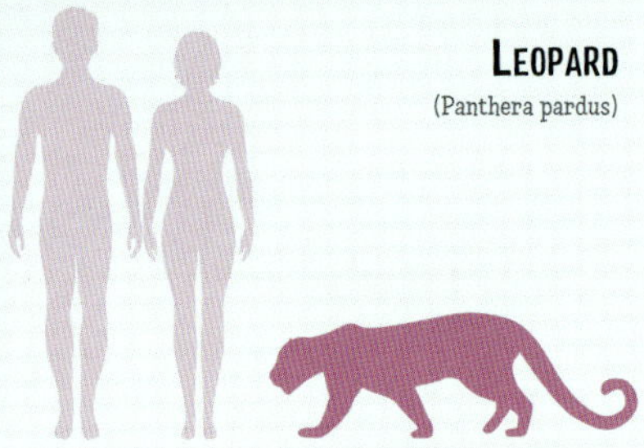

LEOPARD
(Panthera pardus)

SIZE	Up to 2.6 m (8.5 ft) long, including the tail
WEIGHT	Up to 80 kg (175 lb)
GROUP	Leap
HABITAT	Tropical and subtropical grasslands, forests and deserts of Africa and Asia
DIET	Carnivore
CON. STAT.	Vulnerable
LIFESPAN	Up to 15 years in the wild
KEY FACTS	Strong, stealthy hunters; skilled climbers; solitary; nocturnal

TREETOP LARDER Leopards often store their prey in the treetops where it is safe from scavengers, such as hyenas and jackals. Using their immense strength, they haul kills – sometimes twice their own bodyweight – high into the treetops, where they feed on it over several days.

STEALTHY STALKERS

Leopards hunt using a mix of patient, stealthy stalking followed by an explosive final ambush. They are skillful climbers and often position themselves in a tree where they have an excellent view of prey, leaping down on them as they pass below. Leopards kill with a powerful bite to the throat and often bring down prey larger than themselves.

SOUND AND VISION Leopards have exceptionally good eyesight and hearing. They hunt from dusk to dawn and their eyesight is well adapted to low-light conditions. They can see at least six times better in the dark than humans. Their sense of hearing is also acute, and they can rotate their large ears 180 degrees to locate the source of even faint sounds, such as rustling in the undergrowth.

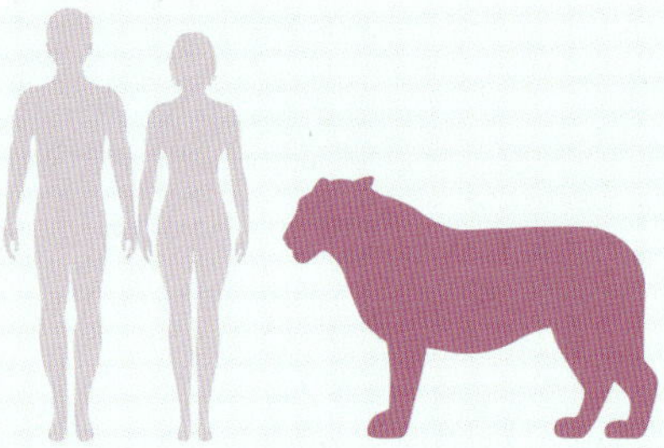

Tiger
(*Panthera tigris*)

SIZE	Up to 2.7 m (8.8 ft) body length + 1 m (3.3 ft) tail
WEIGHT	Up to 420 kg (920 lb)
GROUP	Ambush or streak
HABITAT	Diverse, from cold northern conifer forests to tropical rainforests and mangrove swamps
DIET	Carnivore
CON. STAT.	Endangered
LIFESPAN	About 10-15 years in the wild
KEY FACTS	The largest of the big cats; they love water and are strong swimmers; normally live alone

STRONG SWIMMERS Tigers are skilled swimmers and often chase prey such as deer into lakes and rivers where they have the advantage. While their main diet is deer and wild boar, some tigers also hunt fish, especially if they live in mangrove forests or swamps. Tigers also go in the water to cool down.

STALK AND AMBUSH

Moving silently on its padded paws, the tiger bounds through the snowy forest, eyes firmly fixed on its target. Tigers use stealth to close in on their prey before erupting in a burst of speed or an ambush-leap that puts them right on top of their victim. Their stripy coats help them to blend in with the dappled forests they live in.

THE TIGER FAMILY Tigers all belong to the same species, although they are divided into six subspecies: Siberian, Bengal, Indo-Chinese, South China, Sumatran and Siberian. Tigers are solitary creatures, except for females when they have cubs with them. They normally give birth to a litter of two to four cubs every two years or so. The cubs stay with the mother for 18-24 months.

TREE CLIMBING Tigers can climb trees and they often do in order to see prey better or to escape danger. Young tigers are more likely to climb as fully grown adults can be too heavy. Often they position themselves on a lower branch and wait for prey to pass underneath. Unlike leopards, tigers very rarely haul their prey into the trees to keep it safe from scavengers.

TEAM PLAYERS

Lions are fierce predators that often catch prey much larger than themselves by working together in cooperative groups. They usually hunt at night, stalking prey until they are close enough to ambush it. They use their weight to knock it down and kill it by suffocating it with a bite to the throat or muzzle.

FEMALE HUNTERS While both male and female lions hunt, lionesses usually do 90 per cent of the work as they are lighter and swifter. Often the males will join in when hunting large prey such as buffaloes or when defending the kill, using their greater strength to help. Lions are not always successful when hunting and often resort to scavenging.

SOCIAL CATS Lions are social animals and live together in family groups called prides. They are the only wild cats to do this. The pride is usually based on closely related females, who also hunt cooperatively together. Lionesses are known to nurse and protect each other's cubs and to engage in bonding behaviours like grooming, rubbing heads and roaring together.

Lion

(Panthera leo leo)

SIZE	Up to 1.2 m (4 ft) tall at the shoulders
WEIGHT	Up to 190 kg (418 lb)
GROUP	Pride
HABITAT	Woodlands, grasslands and bush in sub-Saharan Africa
DIET	Carnivore
CON. STAT.	Vulnerable
LIFESPAN	Up to 16 years in the wild
KEY FACTS	Only cats that live in groups; females do most of the hunting

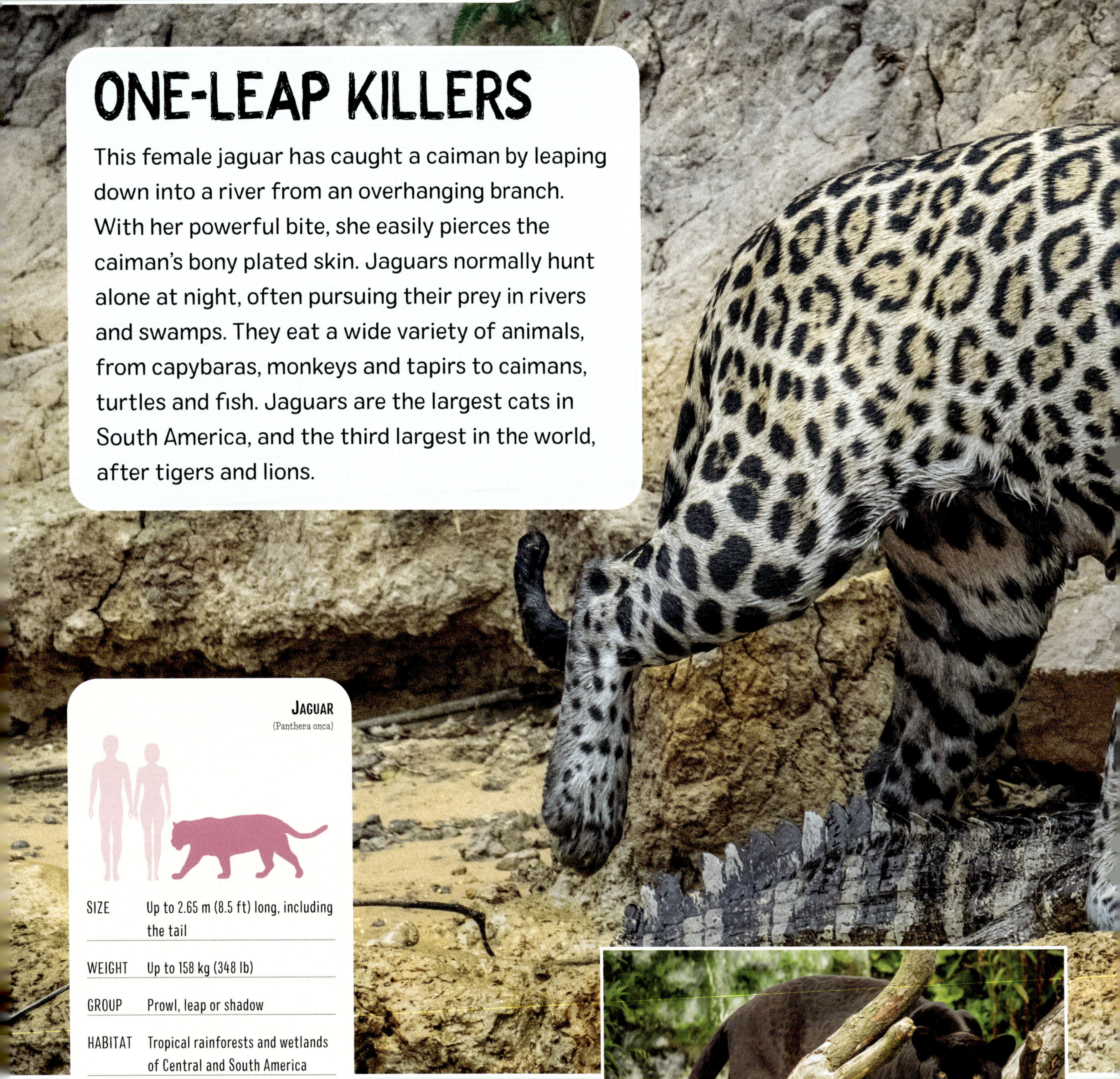

ONE-LEAP KILLERS

This female jaguar has caught a caiman by leaping down into a river from an overhanging branch. With her powerful bite, she easily pierces the caiman's bony plated skin. Jaguars normally hunt alone at night, often pursuing their prey in rivers and swamps. They eat a wide variety of animals, from capybaras, monkeys and tapirs to caimans, turtles and fish. Jaguars are the largest cats in South America, and the third largest in the world, after tigers and lions.

Jaguar

(Panthera onca)

SIZE	Up to 2.65 m (8.5 ft) long, including the tail
WEIGHT	Up to 158 kg (348 lb)
GROUP	Prowl, leap or shadow
HABITAT	Tropical rainforests and wetlands of Central and South America
DIET	Carnivore
CON. STAT.	Near threatened
LIFESPAN	12–15 years in the wild
KEY FACTS	Largest cats in the Americas, powerful bite; good swimmers

BLACK JAGUARS While most jaguars have yellow or gold fur with darker spots, about 10 per cent of them have a black coat that helps them blend in with their dense jungle habitat. They are not a separate species. If you look closely at their coats you can still see their spots.

APTLY NAMED The name 'jaguar' comes from a Native American word that means 'the wild beast that kills with one leap', and neatly captures their hunting style. Sometimes they wait on low-hanging branches, pouncing down on prey up to 7.5 metres (25 feet) below. At others, they jump sideways or straight up in the air to ambush unsuspecting prey.

Jaguars are famous for their incredible ability to make explosive leaps in all directions.

MARINE PREDATORS

Top predators at sea include a variety of shark species, marine mammals like killer whales and seals, and a wide range of fish.

KILLER WHALE
(*Orcinus orca*)

SIZE	Up to 10 m long (32 ft) long
WEIGHT	Up to 10,000 kg (22,000 lb)
GROUP	Pod
HABITAT	Wide-ranging, found in oceans around the world
DIET	Carnivore
CON. STAT.	Uncertain
LIFESPAN	Up to 60–90 years in the wild
KEY FACTS	Highly intelligent social mammals; largest member of the dolphin family; also known as orcas

DIVERSE DIET Killer whales eat a wide range of food, often depending on where they live. They feed on fish and squid like other toothed whales, but will also target seals, sea birds and even whale species much bigger than themselves.

KILLER ON THE BEACH!

Killer whales, also known as orcas, are top ocean predators. They often hunt as a team using highly intelligent, specialised techniques such as coordinated herding and ramming. These strategies are learned within pods and they often divide tasks, with some members acting as drivers and others as attackers. They share food within their pod.

ECHOLATION When hunting in dark or murky waters, killer whales use echolocation to find prey. They emit very high-pitched echolocation clicks, well above the range of human hearing, and can detect fish up to 150 metres (500 feet) away, identifying their size, shape, speed and direction.

Despite their name, killer whales are not technically whales at all; they are the largest member of the dolphin family.

ON THE BEACHES Killer whales in Argentina have been observed driving seals towards the beach where they are slower moving than in the water. The whales follow them out of the water onto the beach to grab them. In Antarctica, killer whales have been seen making waves near ice floes to wash sea lions and seals into the water, where they can capture them.

BOTTLENOSE FEEDERS

Bottlenose dolphins are very intelligent marine mammals known for their acrobatic leaps, ability to communicate and use of tools like sponges to protect their snouts while looking for food. They look so friendly, but we should not forget that they are also highly skilled, top-level predators.

HUNTING TACTICS Bottlenose dolphins often work together to herd schools of fish, such as mullet, towards shallow waters or onto muddy banks, or they simply surround a school of fish in a tight circle, forcing them to huddle together. Then the dolphins take turns diving through the centre of the huddle to feed.

TOOTHY JAWS Bottlenose dolphins have 72 to 116 cone-shaped teeth which they use for grasping prey rather than chewing. They usually swallow their prey whole. Bottlenose dolphins are opportunistic feeders and eat a variety of fish, squid, prawns, crabs and shellfish, depending on what's available. They consume about 13.5 kg (30 lb) of food every day.

BOTTLENOSE DOLPHIN

(*Tursiops erebennus*)

SIZE	2–4 m (6.6–13 ft) long
WEIGHT	Up to 650 kg (1,400 lb)
GROUP	Pod
HABITAT	In oceans all over the world, except for cold polar regions
DIET	Carnivore
CON. STAT.	Least concern
LIFESPAN	40–60 years in the wild
KEY FACTS	Travel in pods of up to 2,000; highly intelligent; communicate with clicks and squeaks

SOCIAL STRUCTURE AND PLAY

Dolphins live together in highly social groups. Individuals within the group form long term bonds that are reinforced by fin rubbing and swimming together. They use high-frequency clicks and whistles for communication. Each dolphin has a unique whistle, which is like a name. Dolphins love to play by leaping out of the water, tossing seaweed and using open-mouth expressions to 'smile'.

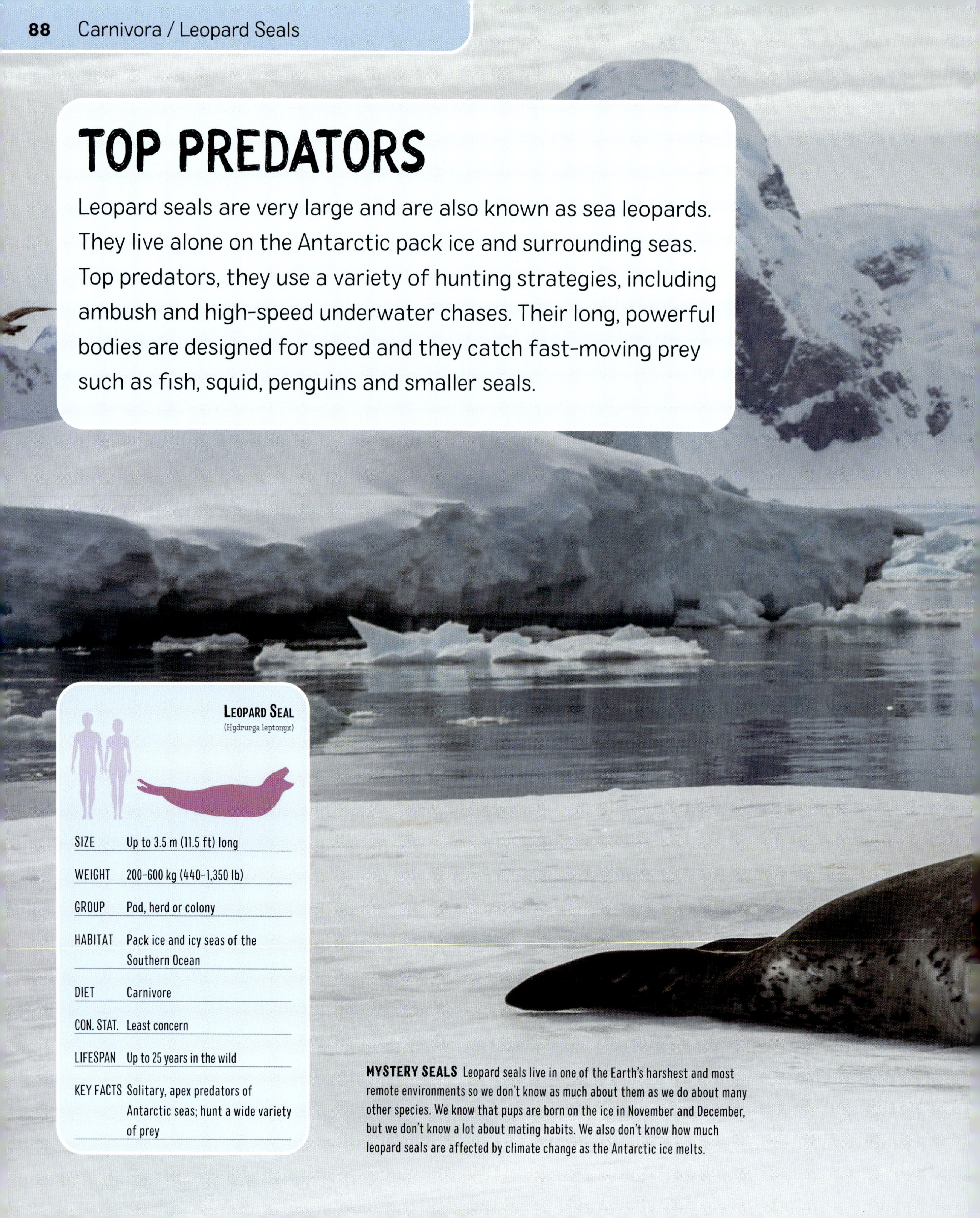

TOP PREDATORS

Leopard seals are very large and are also known as sea leopards. They live alone on the Antarctic pack ice and surrounding seas. Top predators, they use a variety of hunting strategies, including ambush and high-speed underwater chases. Their long, powerful bodies are designed for speed and they catch fast-moving prey such as fish, squid, penguins and smaller seals.

Leopard Seal
(Hydrurga leptonyx)

SIZE	Up to 3.5 m (11.5 ft) long
WEIGHT	200–600 kg (440–1,350 lb)
GROUP	Pod, herd or colony
HABITAT	Pack ice and icy seas of the Southern Ocean
DIET	Carnivore
CON. STAT.	Least concern
LIFESPAN	Up to 25 years in the wild
KEY FACTS	Solitary, apex predators of Antarctic seas; hunt a wide variety of prey

MYSTERY SEALS Leopard seals live in one of the Earth's harshest and most remote environments so we don't know as much about them as we do about many other species. We know that pups are born on the ice in November and December, but we don't know a lot about mating habits. We also don't know how much leopard seals are affected by climate change as the Antarctic ice melts.

SEASONAL EATING Leopard seal diets vary according to the season. During the summer they eat a lot of fish and also use their specially adapted jaws and teeth to filter krill from the seawater. In the winter months they turn to larger prey, such as seals and penguins. Some leopard seals patrol shorelines and ice packs with penguin colonies, waiting for them to dive into the sea. Here we can see a Gentoo penguin racing up a rocky shoreline to escape a leopard seal.

California Sea Lion
(Zalophus californianus)

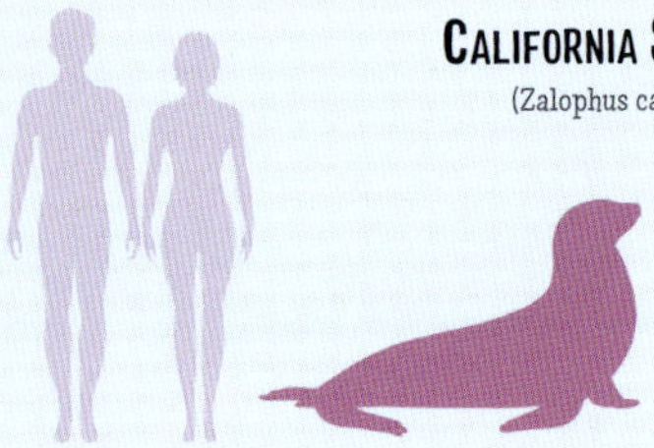

SIZE	Up to 2.4 m (8 ft) long
WEIGHT	Up to 300 kg (660 lb)
GROUP	Colony or raft (at sea)
HABITAT	Coastal waters, rocky shores, and sandy beaches of the North American West Coast
DIET	Carnivore
CON. STAT.	Least concern
LIFESPAN	15–25 years in the wild
KEY FACTS	Intelligent predators; often hunt cooperatively

NICKNAMES Sea lions are often called 'sea dogs' because their faces look a bit like a dog. They also bark to communicate.

SPEEDY AND NOISY Sea lions can swim at 50 km/h (30 mph) and can walk and run on land. They are social animals and live in large groups. They communicate using loud barks, growls, grunts and clicks to maintain social bonds, mark out territories, attract mates and recognise each other.

SEA DOGS

California sea lions are very intelligent, opportunistic feeders that hunt using a mix of speed, agility and specialised senses. They often hunt in groups, herding fish into tight, shallow or rocky areas where they are easier to catch. They also use their sensitive whiskers to detect prey in murky waters and can dive down to 300 metres (1,000 feet) to catch fish and squid. Sea lions are also known to steal fish from commercial and recreational fishing gear.

Walrus

(Odobenus rosmarus)

SIZE	Up to 3.6 m (almost 12 ft) long
WEIGHT	Up to 2,000 kg (4,400 lb)
GROUP	Herd or pod
HABITAT	Sea ice, snow-covered moving pack ice or ice floes in the Arctic
DIET	Carnivore
CON. STAT.	Vulnerable
LIFESPAN	Up to 40 years in the wild
KEY FACTS	Social animals, walruses live in herds; greatest threat is global warming

SENSITIVE WHISKERS Walruses feed twice a day, diving down to the sea floor up to 100 metres (330 feet) below the surface. The water is dark and murky and they use their highly sensitive, stiff whiskers to detect clams and invertebrates buried in the seabed. They shoot a powerful jet of water out of their mouths to expose the food, then use their tongues to suck the soft clams and mussels out of their shells.

FORAGE FEEDERS

Adult walruses weigh about as much as a large family car and need a lot of food just to keep them going. They mainly forage on the sea floor for clams as well as mussels, worms, snails and sea cucumbers. The eat about 45 kg (100 lb) of food every day – that's about 3,000 clams! When food is scarce, walruses also eat fish, seabirds or even the carcasses of young seals.

TOOTHY TOOLS Walruses use their long tusks for a wide variety of tasks, from hauling themselves out of the water to creating breathing holes in the ice. They come in handy as a defensive weapon against polar bears (their only natural predator) and also for establishing dominance within their herds. They don't use their tusks much for digging up food or eating.

TOP PREDATORS

Great white sharks are the largest predatory fish. While an average female great white might grow as long as 4.9 metres (16 feet), some have been measured at six metres (20 feet), or as long as a large delivery van! They are among the fiercest predators in our oceans. They have a superb sense of smell and can swim fast. Great whites make long journeys in search of food and the mild sea temperatures they prefer.

GREAT WHITE SHARK
(Carcharodon carcharias)

SIZE	Up to 6 m (20 ft) long
WEIGHT	Up to 1,900 kg (4,200 lb)
GROUP	Shiver or frenzy (when feeding)
HABITAT	Temperate coastal waters, usually near shorelines with seal colonies
DIET	Carnivore
CON. STAT.	Vulnerable
LIFESPAN	About 70–75 years in the wild
KEY FACTS	Females are larger then the males; highly toxic blood; good camouflage

RENEWABLE GNASHERS Great whites have 28 teeth in the upper jaws and 25 in the lower. But behind the front row, there are always five or six more rows of teeth coming through, each at a different stage of development. The sharks need constantly renewing teeth to replace damaged ones. During their long lives, they can get through up to 30,000 teeth!

INTELLIGENT HUNTERS Great whites can smell blood in the water up to 400 metres (1,300 feet) away. They are ambush predators, usually swimming deep and striking prey from below. They poke their heads out of the water to search for seals and other prey and can launch themselves out of the water in a breach (leap) at over 60 km/h (40 mph). Here you can see a great white making a spectacular leap to catch a seal.

NOT FUSSY EATERS Bull sharks will eat almost anything, although their diet consists mainly of fish, including other sharks. They also sometimes eat rays, dolphins and sea turtles. They usually head-butt prey to stun it before going in for the kill. Their powerful jaws have a tremendous bite force, making them easily strong enough to capture and consume large prey.

Bull Shark

(Carcharhinus leucas)

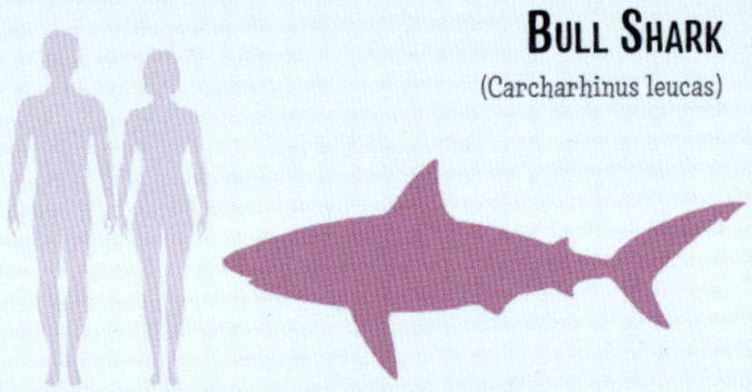

SIZE	Up to 3.5 m (11 ft) long (females)
WEIGHT	Up to 600 kg (1,320 lb)
GROUP	Shiver
HABITAT	Warm, shallow coastal waters and freshwater rivers and lakes around the world
DIET	Carnivore
CON. STAT.	Vulnerable
LIFESPAN	20–30 years in the wild
KEY FACTS	Aggressive predators; unusual ability to live in freshwater habitats

AGGRESSIVE BULLIES

This is the shark that inspired the book that led to Stephen Spielberg's famous 'Jaws' film. Bull sharks are large and one of the top three most dangerous sharks in the world. Bullies are infamous for their unprovoked attacks on swimmers and surfers in murky coastal waters.

BLUNT NOSED The species is named for its short, wide nose, which is thought to resemble that of a bull.

RIVER AND SEA Bull sharks live mainly in warm coastal sea waters but they also have the unique ability to survive for long periods of time in freshwater environments. They often travel far up rivers like the Mississippi and the Amazon and have been known to live in freshwater lakes for years.

ENDANGERED Great hammerheads are critically endangered. Their natural habitat is threatened by coastal development and pollution. What's more, they are slow-growing and mature late, so they cannot reproduce quickly enough to recover from over-fishing. They are also targeted for their large fins, which are used to make a luxury shark fin soup.

Great Hammerhead Shark

(*Sphyrna mokarran*)

SIZE	Up to 6 m long (20 ft) long
WEIGHT	Up to 580 kg (1,280 lb)
GROUP	School
HABITAT	Tropical waters near land all around the world
DIET	Carnivore
CON. STAT.	Critically endangered
LIFESPAN	20–30 years in the wild
KEY FACTS	Bizarre head shape; specialised hunting techniques; in danger of becoming extinct

WEAPONISED HEADS

Hammerhead sharks are famous for their wide T-shaped heads, called cephalofoils. They are specialist predators, with 360-degree vision and thousands of tiny electrical sensors in their heads that help them to hunt for stingrays, fish, squid, octopus and smaller sharks.

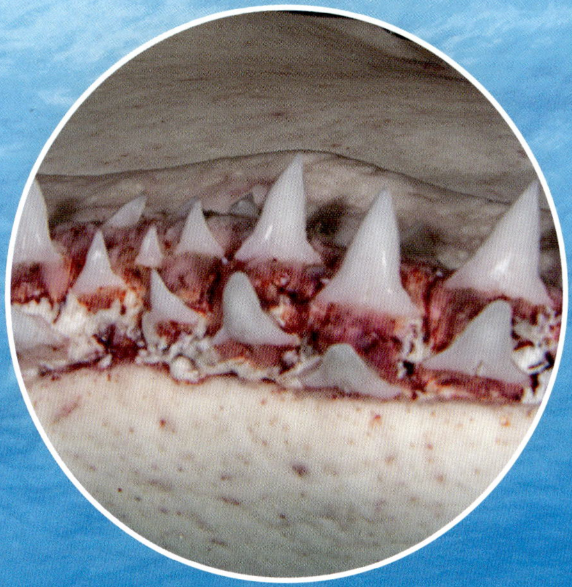

ROWS OF TEETH Hammerheads have about 17 rows of serrated, triangular teeth that are specially designed to grip and slice through the flesh of rays, fish and other sharks. Like all sharks, these teeth continually replace themselves.

HUNTING STRATEGY The hammerhead's very wide head is covered in special electrical sensors and they use it – a bit like we use metal detectors – to find prey buried under the sand. Here you can see a great hammerhead cruising over the sea floor, swaying its head from side to side to increase the sweep. When it locates prey, it often pins it to the seabed with its head.

TRASH CAN DIET

Tiger sharks are apex predators that eat a wide variety of food, including fish, sea turtles, rays, smaller sharks, seabirds, dolphins and crustaceans. Their wide-ranging diet makes them vulnerable to marine pollution, as they often eat the plastic and other toxic rubbish that is accumulating in our oceans. This, combined with overfishing, means that their numbers are declining. The loss of top predators leads to imbalances in the ocean ecosystem.

SOLITARY HUNTERS Tiger sharks usually hunt alone from dusk to dawn. Their eyesight is very good and they have a special reflective layer in their eyes that acts like a mirror to amplify the dim light. They are able to see in the dark about ten times better than humans.

SQUARE JAWS AND STRIPES Tiger sharks have almost square jaws that are full of special, serrated teeth designed to cut through turtle shells and other tough prey. The species is named for the dark, vertical stripes and spots found on their bodies. These are most visible in juveniles and fade as the sharks grow up.

OPPORTUNISTS Tiger sharks are opportunist feeders, which means that they will eat anything they can lay their teeth on. This one is about to gobble up an unsuspecting albatross.

LARGER FEMALES Tiger sharks are among the largest shark species. The females are much bigger and heavier than the males, sometimes reaching seven metres (23 feet) and weighing over 1,500 kg (3,300 lb). Female tiger sharks give birth to live young and they need the extra body size to carry their pups safely.

TIGER SHARK
(*Galeocerdo cuvier*)

SIZE	Up to 7 m (24 ft) long
WEIGHT	Up to 1,500 kg (3,330 lb)
GROUP	Shiver or school
HABITAT	Warm coastal seas and river estuaries around the world
DIET	Carnivore
CON. STAT.	Near threatened
LIFESPAN	30–50 years in the wild
KEY FACTS	Apex predators that eat a wide range of food

BOTTOM FEEDERS

Spotted eagle rays glide over the ocean floor using special receptors in their sensitive snouts to locate prey such as crabs, clams and small molluscs buried in the sand. They use their snouts to root them out and gobble them up.

SPOTTED EAGLE RAY

(Aetobatus narinari)

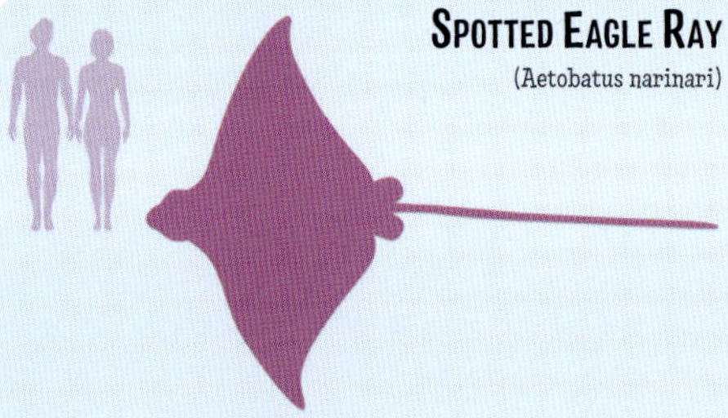

SIZE	About 5 m (16 ft) long, including their tails
WEIGHT	Up to 230 kg (500 lb)
WINGSPAN	Up to 3 m (10 ft) wide
GROUP	Fever
HABITAT	Tropical waters in the Atlantic and Indo-Pacific Oceans
DIET	Carnivores
CON. STAT.	Near threatened
LIFESPAN	15–25 years in the wild
KEY FACTS	Use sensitive noses to find food; very large; mouths on the bottom of snouts

ELECTRO-SENSORY NOSES The ray's rostrum (snout) is full of electro-sensory pores that act like a radar to find prey by detecting their electrical fields. Their shovel-shaped snouts are perfect for digging them up. They can even use them to turn a stone over. The ray's mouth is under the snout and full of flattened teeth that work like grinders to crush their hard-shelled prey,

DIGGERS The rays raise a cloud of sediment and sand which they flush through their gills with the seawater. They are often followed by smaller fish, such as jacks, kingfish and trevally, that feed on any of the food that the stingray misses.

GORGEOUS SPOTS The rays' backs are covered in striking dark blue or blackish patterns that help them to blend in with their surroundings when seen from above. Sometimes they lie on the sea floor, invisible to predators. Like the stripes on a tiger, the patterns on each ray are unique and help other rays to recognise them.

PACIFIC PREDATOR

With an arm span of up to nine metres (30 feet) and weighing as much as 270 kg (600 lb), the GPO (giant Pacific octopus) is the largest octopus in the world. Highly intelligent, it is an apex predator in its marine environment, using a deadly combination of stealth, camouflage and venom to hunt clams, crabs, lobsters, other octopuses and fish.

MASTERS OF DISGUISE Like most octopuses, GPOs can quickly change their skin's colour, pattern and texture using specialised cells called chromatophores. This allows them to blend in with their surroundings so they can sneak up on prey undetected.

POWERFUL SUCKERS

GPOs have more than 2,000 powerful suckers on their eight arms and they use them to grip, taste and smell their world. When hunting, they use their suckers to grasp prey, holding them still while they deliver a shot of venom through their sharp beak, which paralyses their victims until they are ready to eat them.

Giant Pacific Octopus
(*Enteroctopus dofleini*)

WEIGHT	Up to 270 kg (600 lb)
ARM SPAN	Up to 9 m (30 ft)
GROUP	Consortium or tangle
HABITAT	In tide pools and coastal waters of the North Pacific Ocean
DIET	Carnivore
CON. STAT.	Least concern
LIFESPAN	3–5 years in the wild
KEY FACTS	Highly intelligent; apex predators that help balance food webs; use ink clouds and camouflage to hunt and avoid being eaten

BEAUTIFUL BUT DEADLY

Blue-ringed octopuses are incredibly beautiful, but if you are lucky enough to see one in the wild you should never pick it up. The saliva of these small octopuses contains two deadly types of venom and a single bite is enough to kill. The octopuses use one venom to immobilise prey while they eat it, and the other to defend themselves against predators such as eels, large fish and dolphins.

SEVERAL SPECIES There are at least four different species of blue-ringed octopus. They are all equally venomous and can change colour to warn off predators and blend into their surroundings. They live in burrows in the sea floor and only come out to hunt or search for a mate.

GREATER BLUE-RINGED OCTOPUS
(Hapalochlaena lunulata)

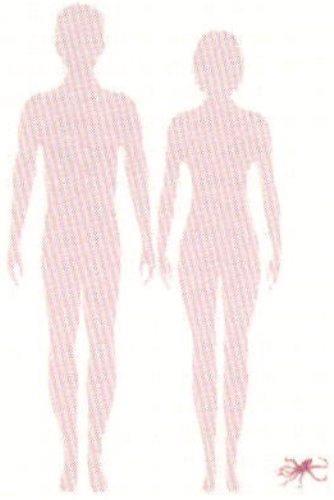

SIZE	Up to 20 cm (8 in) long
WEIGHT	Up to 90 g (3 oz)
GROUP	Tangle, consortium
HABITAT	Shallow warm waters of the Pacific and Indian oceans
DIET	Carnivore
CON. STAT.	Least concern
LIFESPAN	About 2–3 years in the wild
KEY FACTS	Highly venomous; makes its blue rings flash when threatened

HUNTING TECHNIQUES Blue-ringed octopuses are ambush predators, hiding in burrows, crevices and empty seashells (or discarded bottles), waiting for prey to come by before pouncing and trapping it with their arms. They use their sharp, parrot-like beaks to pierce the hard shells of their prey and inject a venom containing a potent toxin that paralyses them instantly.

COLOUR MASTERS Blue-ringed octopuses have about 60 blue rings spread evenly over their bodies. Under the rings their skin colour varies from yellow to ochre or white. When they are threatened, they can flash their rings a bright iridescent blue as a warning sign: it means 'Stay away! I'm deadly poisonous!'

COLOURFUL HUNTERS

With their bright blue backs, silvery-white undersides and sleek, aerodynamic bodies, blue marlin are magnificent fish. They are large, with the females almost four times the size of the males. They are also fast and make long, solitary migratory journeys. Blue marlin are opportunistic hunters, feeding mainly on fish and squid.

Blue Marlin
(*Makaira nigricans*)

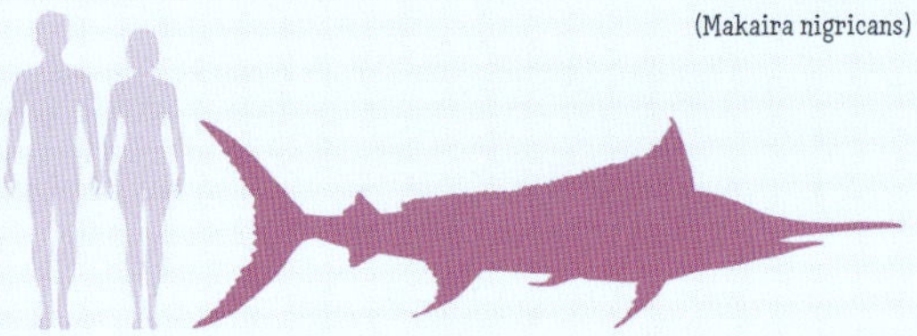

SIZE	Up to 5 m (16 ft) long (females)
WEIGHT	Up to 900 kg (2,000 lb)
GROUP	School or shoal
HABITAT	Warm tropical waters of the Atlantic, Pacific and Indian Oceans
DIET	Carnivore
CON. STAT.	Vulnerable
LIFESPAN	27–30 years in the wild (females live longest)
KEY FACTS	Large, fast ocean-going fish; females are much larger than the males; use sophisticated hunting tactics like herding

FAST AND FIERCE Blue marlins are among the fastest fish in the ocean. They hunt in surface waters, often locating prey by following feeding birds. They sometimes work as a team, using their long, spear-like bills to herd schools of small fish into tight, dense balls. Then they swim through the ball at high speeds, using their sharp bills to slash, stun and kill the fish.

WHEN THE HUNTERS ARE HUNTED There are four main species of marlin: Blue, Black, Striped and White. They are all considered prime game-fish and are pursued in offshore waters all over the world for sport rather than commercial food consumption. This black marlin has just been hooked.

Green Moray Eel
(*Gymnothorax funebris*)

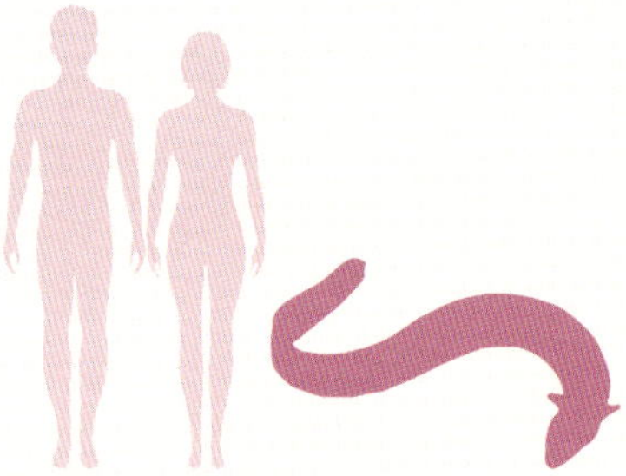

SIZE	Up to 2.7–3 m (9–10 ft) long
WEIGHT	Up to 35 kg (78 lb)
GROUP	Bed or swarm
HABITAT	Warm, shallow coastal waters, including coral reefs, mangrove forests and rocky shorelines
DIET	Carnivore
CON. STAT.	Least concern
LIFESPAN	Up to 10–30 years in the wild
KEY FACTS	One of the largest species of moray eel; has a second set of jaws in its throat

NIGHT HUNTERS

Green moray eels poke their heads out of rocky crevices during the day, but they only come out to hunt at dusk. Although they have poor eyesight, they have a highly developed sense of smell which helps them to locate prey in the dark. They are ambush predators and mainly hunt fish, crabs, squid and octopuses.

TEAM PLAYERS Moray eels have been observed hunting cooperatively with groupers (large coral reef fish). When a small fish escapes from a grouper and zips into a crevice to hide, the grouper is too large to follow. It goes to a nearby moray eel and does a dance, asking the eel to follow it. They go to the crevice and the eel goes in to grab the fish. If it misses, the grouper catches it on its way out!

BROWN NOT GREEN Moray eels are actually brown. They look green because their skin is covered with a thick layer of yellow mucus to protect them from parasites and bacteria. Underwater, this yellow mucus makes them appear green.

TWO SETS OF JAWS Moray eels have a unique, second set of jaws in their throats, called pharyngeal jaws. They can move these forwards into their mouths to grab prey and drag it directly into their throats. Moray eels don't have limbs to manage food and this adaptation allows them to swallow large prey whole.

DEATH BALLOONS

There are over 150 species of pufferfish. The smallest is just 2.5 cm (1 inch) in length, while the largest is over 70 cm (28 inches) long. They are all predatory carnivores with excellent eyesight and fierce ambush tactics. They have a special set of fused teeth and strong jaw muscles to crush the shells of the crabs, molluscs and snails they feed on.

PUFFERFISH
(Tetraodontidae)

SIZE	Up to 70 cm (2 ft) long
GROUP	School or shoal
HABITAT	Warm, tropical waters around the world, usually in coastal areas, coral reefs and seagrass beds
DIET	Carnivore
CON. STAT.	Least concern (most species; some are endangered)
LIFESPAN	5–10 years in the wild
KEY FACTS	Small to medium predatory fish; usually very toxic; unique ability to puff up for self defense

INFLATION The most amazing thing about pufferfish is that they can puff themselves up into balls that are two to four times their normal size. They do this by pumping water or air into a special chamber inside their stomachs. This enables them to defend themselves from predators.

TOXIC DEATH BALLS Almost all pufferfish are highly toxic. Just one pufferfish has enough toxin to kill 30 adult humans. They are not born with the toxin; they accumulate it over time from the food the eat. In Japan, skilled chefs prepare the non-toxic parts that are considered a great delicacy.

DEADLY STINGERS

Despite their delicate appearance, sea nettle jellyfish are top predators in many habitats. Their long, trailing tentacles are covered with stingers. They hunt by drifting through the water, using their tentacles to sting and stun prey. Then they use special frilly structures, called oral arms, to bring the food to their mouths at the centre of the bell.

Sea Nettle Jellyfish
(*Chrysaora*)

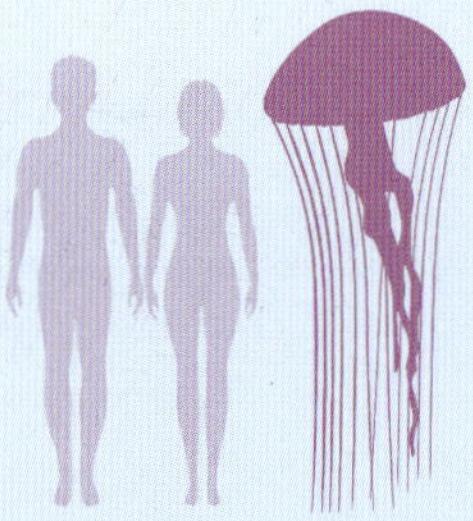

SIZE	15–100 cm (6–39 in) bell diameter; 1.5–6 m (5–20 ft) trailing tentacles
GROUP	Smack, bloom or swarm
HABITAT	Warm to temperate coastal waters around the world
DIET	Carnivore
CON. STAT.	Least concern
LIFESPAN	6–12 months in the wild
KEY FACTS	They have no brain, no heart and no blood; some species grow very large

MANY SPECIES There are 15 or 16 different species of sea nettles living in various oceans of the world. They are one of the most common jellyfish found in coastal waters worldwide, from the Atlantic and Pacific oceans to the Mediterranean Sea. They are famous for sometimes appearing in massive 'blooms' made up of thousands of individual jellyfish. These swarms usually happen when sea temperatures rise or when there is a lot of pollution.

DON'T GET STUNG! Sea nettle jellyfish can inflict painful stings on humans. Their venom is designed to paralyse small prey and typically only causes mild to moderate pain lasting 10–20 minutes in humans.

RAM BITERS

Great barracudas are aggressive ambush predators. Blending in perfectly with their surroundings, they lie in wait, using their keen vision to search for passing prey. When it appears, they attack so quickly that their victim doesn't even see them coming. Holding it in their fearsome jaws, they bite repeatedly in quick succession, shaking their heads and slicing the prey in half with their razor-sharp teeth. This is known as ram-biting.

GROUP HUNTING Young great barracudas and other smaller species of barracuda often form large schools to hunt cooperatively. Working together, they herd prey such as sardines or anchovies, forcing them into shallow water or tight 'bait balls', then they attack.

NEVER EAT A BARRACUDA! Barracudas are poisonous and not fit for human consumption. As they age, they accumulate toxins from the smaller fish they eat, which in turn feed on toxic algae. The toxin is not destroyed by cooking, freezing or marinating.

TORPEDO-SHAPED The barracuda's slender cylindrical body is built for speed. The lower jaw protrudes and the razor-sharp teeth face both backwards and forwards to help the fearsome predator grip its slippery prey.

Great Barracuda
(*Sphyraena barracuda*)

SIZE	Up to 2 m (6.5 ft) long
WEIGHT	Up to 50 kg (110 lb)
GROUP	Battery, barrage
HABITAT	Warm coastal waters, coral reefs, sea grass meadows and mangrove forests
DIET	Carnivore
CON. STAT.	Least concern
LIFESPAN	14–18 years in the wild
KEY FACTS	Fierce predators; built for speed; toxic to humans

LONG HISTORY Barracudas have been around for at least 50 million years. While they are not endangered, they are affected by damage to coral reefs, mangrove forests and seagrass beds. They are also caught as bycatch (unintentional capture by commercial fishing).

DEVIL FISH

The lionfish's bizarre appearance and voracious appetite have earned it several nicknames, including 'devil fish' and 'Hoover fish'. They eat over 70 different species of fish and invertebrates. Lionfish hunt alone or cooperatively in groups from dusk to dawn by moving slowly to get close to prey. They sometimes use their spines and fins to herd small fish, or blow water at them to confuse them. When close enough, they strike rapidly, sucking the prey into their mouths.

LIONFISH YAWN Lionfish often swim or hover with their mouths wide open. This increases the amount of oxygen their gills can extract from the water. It also helps with their hunting tactics of blowing water at prey and striking rapidly.

THE SPINES Lionfish have 18 venomous spines, 13 of which stand tall along their backs. The spines are mainly used for defense and the venom is potent enough to kill or paralyse smaller fish.

Common / Red Lionfish

(Pterois volitans / miles)

SIZE	Up to 47 cm (18.5 in) long
WEIGHT	Up to 1.1 kg (2.4 lb)
GROUP	Harem
HABITAT	Warm, tropical waters of the Indo-Pacific Ocean, from Japan and Australia to the Red Sea
DIET	Carnivore
CON. STAT.	Near threatened
LIFESPAN	10–15 years in the wild
KEY FACTS	Skilled hunters blow water at prey; covered in venomous spines

CAMOUFLAGE Common and red lionfish have bold stripes and long fins and spines all over their bodies. This helps them to blend in with the coral, sponges and shadowy crevices in their habitats. They can also change the overall colour of their bodies, becoming darker to match shadowy, reef crevices or lighter for open, sunlit areas.

AIRBORNE PREDATORS

Sky predators include birds of prey such as eagles, falcons and owls, as well as meat-eating bats that use echolocation to catch insects, frogs and fish. Insects like dragonflies and ladybirds are also voracious predators.

STAYING SAFE Adult great horned owls are apex predators and have few natural enemies, but their eggs and fledglings are tempting treats for raccoons, foxes, crows, ravens and other birds of prey. Female owls use piercing shrieks and trills, loud barks, harsh squawks, hisses and bill snapping to keep hungry predators away from the nest. Meanwhile, the male stands guard, letting his partner know when danger approaches.

QUIET STEALTH Great horned owls are famous for their soundless flight. Their soft, velvety feathers break up air flow and absorb sound so they can get very close to prey before they even realise danger is near. The owls are guided by their exceptional hearing and fly slowly just above the ground, surprising victims from above.

Sometimes they sit silently on high branches or poles, using their speckled feathers as camouflage, and dive-bomb passing prey. They also hunt on foot for small prey or even wade into the water to catch frogs and fish.

DON'T MESS WITH ME

Wings up and talons out, this great horned owl mum is vigorously defending her nest. Female owls are 10–20 per cent larger than males and the extra heft is used to cover more eggs and protect the family nest and owlets. Great horned owls mate for life. They divide the care of their young, with the smaller, swifter male hunting and bringing food.

Great Horned Owl
(Bubo Virginianus)

SIZE	45–65 cm (18–26 in) long
WEIGHT	1–2.5 kg (2–5.5 lb)
WINGSPAN	1–1.5 m (3.3–4.8 ft
GROUP	Parliament
HABITAT	Diverse, all over the Americas, from tundra to rainforests, including deserts, grasslands, swamps and city parks
DIET	Carnivore
CON. STAT.	Least concern
LIFESPAN	5–15 years in the wild

SILENT ASSASSINS

Tawny owls are nocturnal, which means they are active at night. These owls don't use echolocation to hunt in the dark, relying instead on special low-light vision and acute hearing to find prey. They also have specialised wing feathers with soft, fringed edges, so they fly in almost total silence.

TERRITORIAL PAIRS Tawny owls are highly territorial. Young owls mate for life and the pair defends a 12–20 hectare (30–50 acre) territory year round. They hunt a variety of prey, from mice, voles and rabbits to frogs, earthworms, beetles and bats.

NESTING Tawny owls like to nest in tree cavities. They also use other natural sites, such as old crow or pigeon nests, squirrel drays, the forks of tree trunks and even rabbit holes. Females lay two to three eggs in the spring which they incubate for around a month. The baby owls are fully independent after about three months.

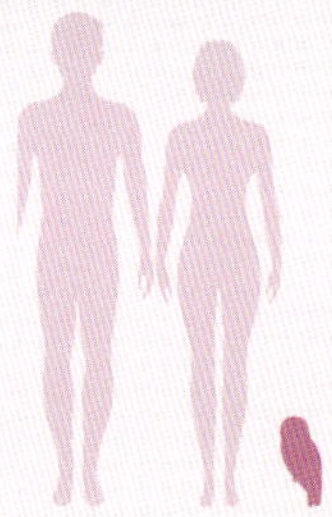

TAWNY OWL
(Strix aluco)

SIZE	Up to 40 cm (16 in) body length
WEIGHT	Up to about 500 g (1.1 lb)
WINGSPAN	Up to 104 cm (41 in)
GROUP	Parliament
HABITAT	Deciduous woodlands, farmland, parks and gardens in Europe and parts of Asia
DIET	Carnivore
CON. STAT.	Least concern
LIFESPAN	4–5 years in the wild
KEY FACTS	Uses specialised eyesight and hearing to hunt; can turn head 270 degrees; 'twit-twoo' duet call

BLENDING IN Tawny owls have mottled brown and grey plumage and blend in perfectly with tree bark, making them almost invisible while roosting during the day.

SHARP TALONS AND SWIVEL TOES Tawny owls use their strong feet and sharp talons (claws) to grasp prey. Like all owls, they have a flexible outer toe that can swivel backwards, allowing them to have two toes pointing forwards and two backwards to grasp prey firmly.

A WAITING GAME

The snowy owl, also known as the Arctic owl, is one of the largest owls in the world. Unlike most owls, they hunt during the day. They use a sit-and-wait technique of perching on high vantage points such as trees or fence posts for hours, using their superb vision and hearing to detect prey, often at a distance or buried under the snow.

BREEDING Snowy owls are fiercely territorial and, if they feel threatened or if their nest is threatened, they will dive-bomb anyone who approaches. Females lay 3–12 eggs – depending on how much food is available – and both adult birds help raise the chicks.

LEMMINGS DIET Snowy owls are not picky eaters, adapting to whatever is available in their habitat to meet their high energy needs. During the summer, snowy owls feed mainly on lemmings, eating three to five every day. But in wintertime they switch to a more varied diet of birds and larger mammals.

TABLE MANNERS Snowy owls swallow their food whole or in large chunks, and all the fur, feathers and bones are compressed into a fur ball called a pellet, which they spit up daily.

IN-BUILT SNOWSHOES Snowy owls have thickly-feathered legs and feet to insulate them from the extreme Arctic cold. The feathers go all the way down to their talons, like a thick, fluffy pair of trousers.

Snowy owl
(Bubo scandiaca)

SIZE	52–71 cm (20–28 in) long
WEIGHT	1.7–3 kg (3.7–6.6 lb)
WINGSPAN	125–170 cm (4–5.5 ft)
GROUP	Parliament or blizzard
HABITAT	Arctic tundra of North America, Europe and Asia
DIET	Carnivore
LIFESPAN	9–10 years in the wild
CON. STAT.	Vulnerable
KEY FACTS	One of the largest owls in the world; daylight hunters; superb hearing and vision

DIVE BOMBERS

The little blue kingfisher perches patiently on a branch about two metres (seven feet) above the river. Suddenly, it spies a fish! It bobs its head slightly to estimate its depth, then plunges into the water, scooping the fish up in its beak. It returns to its perch where it bashes the fish against the wood to snap its spine, then swallows it down, head first.

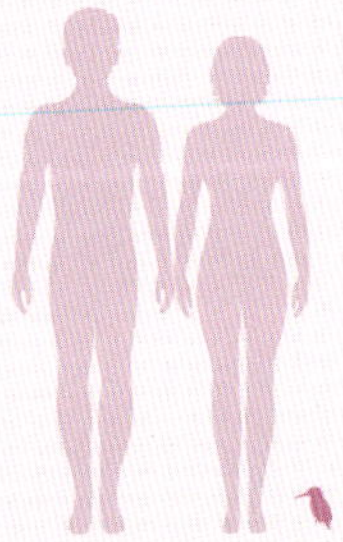

Common Kingfisher
(Alcedo atthis)

SIZE	Up to 16 cm (6.5 in) long
WEIGHT	35-45 g (1.2-1.6 oz)
WINGSPAN	Up to 25 cm (10 in)
GROUP	Concentration, crown or realm
HABITAT	Eurasian and North African rivers and lakes
DIET	Carnivore
CON. STAT.	Least concern
LIFESPAN	5-15 years in the wild
KEY FACTS	Superb divers, great eyesight and dagger-like beaks

PRECISION DIVING Kingfishers always dive head first, often hitting the surface of the river at 40 km/h (25 mph)! Their excellent vision allows them to judge distance and depth, making them accurate even at steep angles. They have a transparent third eyelid that shuts as they enter the water, protecting their eyes while allowing them to continue tracking their prey.

KINGLY COURTSHIP Male kingfishers offer gifts of fresh fish and tasty insects to their sweethearts. Instead of building a nest, common kingfishers dig tunnels into sandy riverbanks and lay their eggs in a chamber at the end. Both parents work to prepare the nest and raise their chicks.

BIG APPETITES To survive, kingfishers have to eat more than half their bodyweight in food every day. A pair with a nest full of chicks may catch over 120 fish per day.

NORTHERN GANNETS
(Morus bassanus)

SIZE	Up to 1 m (3.3 ft) long
WEIGHT	2.3–3.6 kg (5–8 lb)
WINGSPAN	Up to 1.8 m (6 ft)
GROUP	Gannetry or plunge
HABITAT	North Atlantic ocean
DIET	Carnivore
CON. STAT.	Least concern
LIFESPAN	Around 17–20 years in the wild
KEY FACTS	Largest bird in the North Atlantic; spectacular, high-speed divers; underwater vision

TORPEDO DIVERS Gannets fold their wings back like a torpedo right before hitting the water, diving deep to catch their prey.

SOUTHERN GANNETS Northern gannets are closely related to Australasian gannets, and may even be part of a single super-species. Both birds nest in vast, crowded colonies on sea cliffs or rocky, isolated islands. They build cup-shaped nests of seaweed, grass and mud.

PLUNGE DIVERS

A group of northern gannets soars above the roaring ocean, scanning the surface with their exceptional binocular vision to spot prey. All of a sudden, folding their wings out behind them, they plunge-dive into the waves, hitting the water at speeds up to 96 km/h (60 mph). They have air sacs to cushion the impact, excellent underwater vision, and can dive up to 22 metres (72 feet) deep.

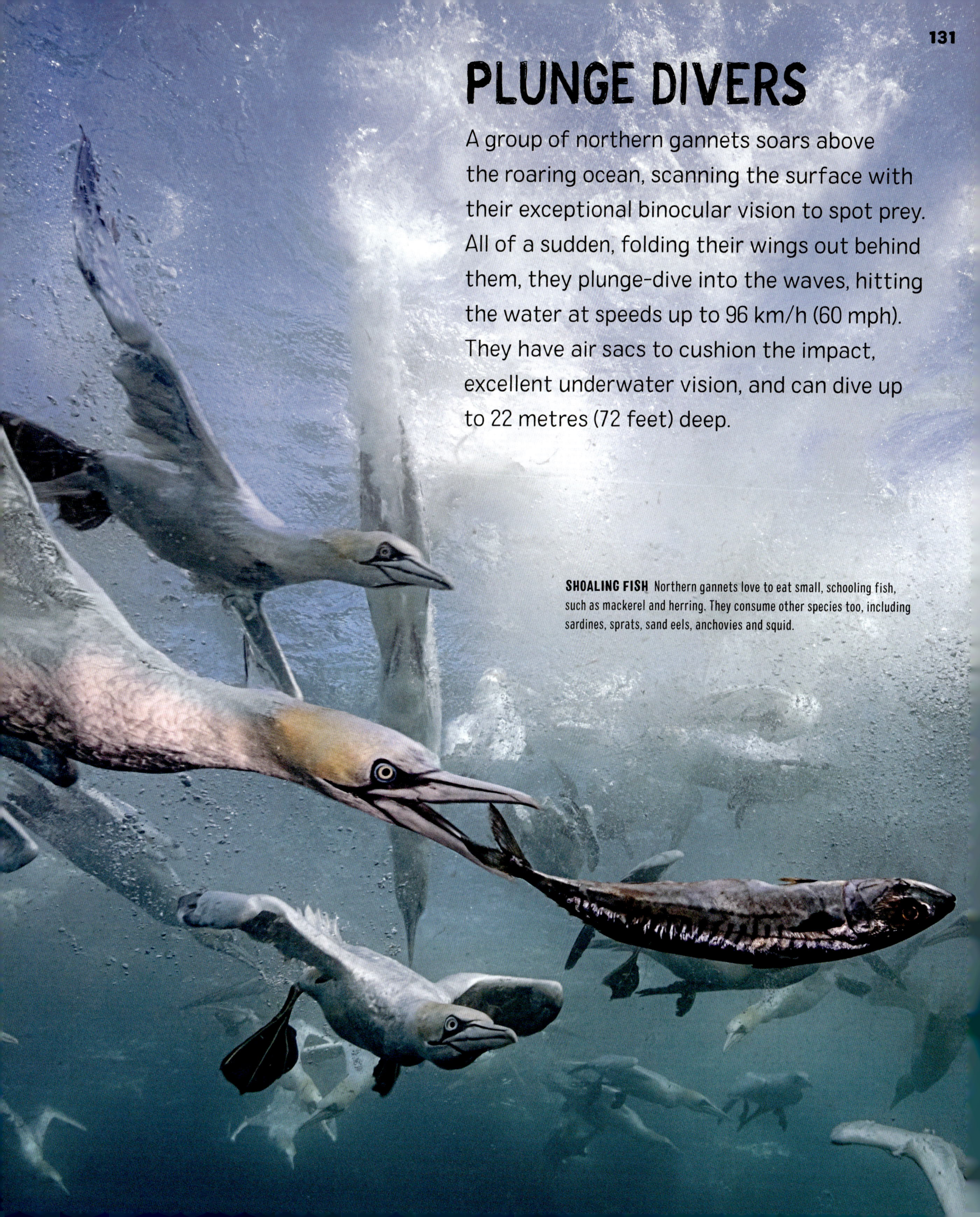

SHOALING FISH Northern gannets love to eat small, schooling fish, such as mackerel and herring. They consume other species too, including sardines, sprats, sand eels, anchovies and squid.

Bald Eagle

(*Haliaeetus leucocephalus*)

SIZE	70–102 cm (28–40 in) long
WEIGHT	3–6.3 kg (6.5–13.8 lb)
WINGSPAN	1.8–2.3 m (5.11–7.7 ft)
GROUP	Convocation
HABITAT	North America, in wetlands such as seacoasts, rivers, large lakes or marshes and other large areas of open water with plenty of fish
DIET	Carnivore
CON. STAT.	Least concern
LIFESPAN	20–30 years in the wild
KEY FACTS	Very keen eyesight; mate for life; are not bald!

CONSERVATION SUCCESS STORY In the 1960s, bald eagles were on the brink of extinction. From half a million birds, the population in the US had dropped to just 400 nesting pairs, mainly due to the widespread use of a pesticide called DDT. The government and wildlife groups stepped in, banning DDT and reintroducing the eagles to their historic habitats. Today, there are over 300,000 bald eagles in the United States,

EAGLE-EYED

Bald eagles mainly eat fish. They soar high above the waves, using their incredible eyesight to spot fish up to a 1.6 km (1 mile) away. When they attack, they swoop down at up to 160 km (100 miles) an hour! They snatch the fish with their sharp, powerful talons, often without breaking the surface of the water, although they can dive when necessary.

HUGE NESTS After some spectacular courtship rituals, bald eagles mate for life. Together each pair builds a large aerie (nest), returning to it year after year. Every year they restore the nest, making it larger. An average bald eagle nest is 1.5–1.8 metres (5–6 feet) in diameter! The female lays one to three eggs per breeding season and both parents care for the young (known as eaglets or chicks) which hatch after about 35 days of incubation and usually leave the nest at around 10–14 weeks old.

LETHAL STOMPERS

Secretary birds are unique in many ways. They are one of the few birds of prey that hunt on foot. Even though they are skillful fliers and often take to the air, they prefer to spend most of their time on the ground. Snakes are their favourite food, and they stomp them to death before eating.

BLACK CREST Secretary birds have a crest of long black feathers on the back of their heads. They look like old-fashioned quill pens that secretaries would use and – according to some – this is how the species got its name.

LONG LEGS Secretary birds have the longest legs of any bird of prey. They walk about 20–30 km (15–20 miles) a day. When chasing prey, they can reach speeds of up to 30 km/h (20 mph). Adult pairs mate for life and raise their chicks together.

SECRETARY BIRDS
(*Sagittarius serpentarius*)

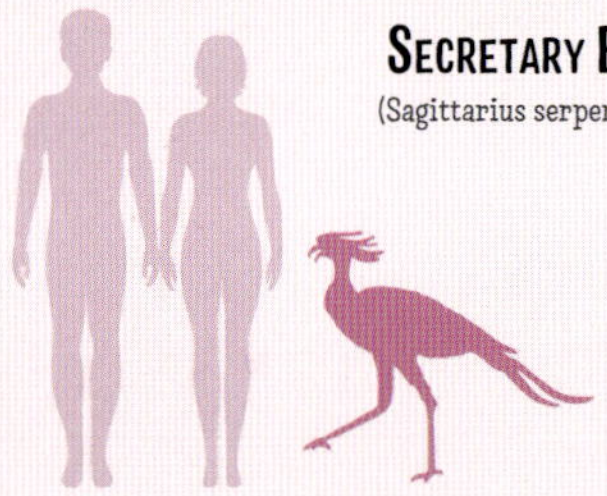

SIZE	Up to 1.5 m (5 ft) tall
WEIGHT	Up to 4.25 kg (9.4 lb)
WINGSPAN	Up to 2.1 m (7 ft)
GROUP	Flock, committee or office
HABITAT	African savannas and grasslands
DIET	Carnivore
CON. STAT.	Least concern
LIFESPAN	10–15 years in the wild
KEY FACTS	One of the few birds of prey that hunt on foot; unique snake-stomping hunting technique; very long legs

STOMP STOMP STOMP Secretary birds hunt in small groups or with a partner from dawn until dusk. When they find a snake, they use their large feet and sharp claws to stomp it to death. They have thick scales on their lower legs to protect them from snake bites. Secretary birds also eat insects, spiders and beetles.

BICYCLE SHORTS The black feathers at the top of its legs make the secretary bird look like it's wearing bicycle shorts!

BONE EATERS

Bearded vultures are very large birds of prey that feed almost exclusively on the bones of already-dead animals. About 80–90 per cent of their diet consists of bone marrow. They are the only birds we know of that feed in this manner.

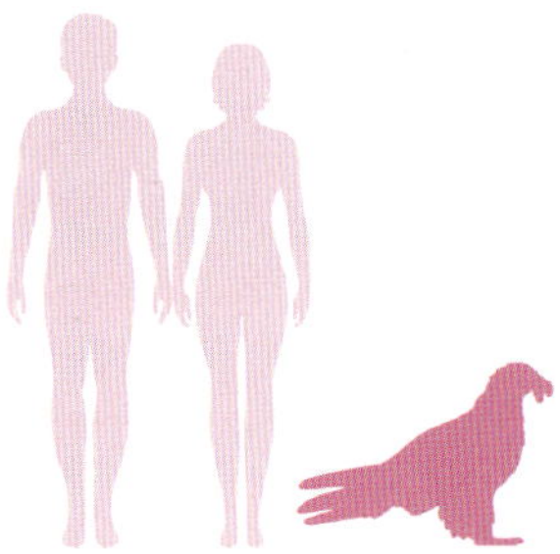

BEARDED VULTURE
(Gypaetus barbatus)

SIZE	Up to 1.25 m (4 ft) long
WEIGHT	4.5–8 kg (10–18 lb)
WINGSPAN	Up to 3 m (10 ft)
GROUP	Committee, kettle or wake
HABITAT	High mountains, such as the Pyrenees, Alps, Himalayas and other mountainous regions of Africa and Asia
DIET	Carnivore
CON. STAT.	Near threatened
LIFESPAN	30–40 years in the wild
KEY FACTS	Specialised diet of bones; huge wingspans

DYED FEATHERS Bearded vultures have white breast, neck and head feathers but the birds dye them a reddish-brown colour by bathing in iron-rich mud. Naturalists think they do this to make themselves look more attractive to prospective mates or to gain status within their group. Some others believe that the mud might have anti-bacterial or anti-parasitic properties, so it makes them feel better.

BONE SWALLOWERS Bearded vultures swallow smaller bones whole. For larger ones, they drop them onto rocks from high above to smash them into bite-sized pieces. They have a specialised digestive system that breaks them down within 24 hours.

REPRODUCTION Bearded vultures live alone except during the breeding season, when they form strong, long-term pair bonds. They nest on remote, high cliffs and raise one to two chicks together.

LOSING GROUND The population of ornate hawk-eagles is declining rapidly mainly because of habitat loss and hunting. They need large, unbroken tracts of tropical forest for survival and these are being destroyed by logging and farming. As their forest homes are cleared, especially in the Amazon, they lose critical nesting and hunting grounds. The birds are also often shot by people who fear they will eat their chickens and other livestock.

DEATH FROM ABOVE

This ornate hawk-eagle is perched on a branch in the upper canopy of a rainforest in Belize. It peers into the dense foliage beneath its feet, waiting to spot a monkey, bird or reptile. Suddenly, it drops from its perch, diving through the thick leaves to grab its prey in its sharp talons.

ORNATE HAWK-EAGLE
(*Spizaetus ornatus*)

SIZE	56–68 cm (22–27 in) long
WINGSPAN	Up to 1.5 m (5 ft)
WEIGHT	Up to 1.8 kg (4 lb)
GROUP	Congress or convocation
HABITAT	Dense, humid tropical rainforests, from southern Mexico to northern Argentina
DIET	Carnivore
CON. STAT.	Near threatened
LIFESPAN	Up to 15 years in the wild
KEY FACTS	Strong, can capture prey up to twice its size; head crest

BLACK CRESTS Ornate hawk eagles are famous for their crest of tall black feathers, which they raise whenever they are excited.

CAREFUL PARENTS Ornate hawk-eagles usually only breed every other year. The parents work together to build a large nest made out sticks about 20–30 metres (65 to 100 feet) above the forest floor. After courtship, females lay a single large egg in the nest, then they incubate it until it hatches. Both parents feed and protect the chick, which stays with them until it is more than a year old.

An adult ornate hawk eagle carries a large stick back to the tree where it is building a nest.

GIVE IT BACK!

Steller's sea eagles are large, aggressive raptors. They often steal each other's fish and then fight to get it back. They are the world's heaviest eagles and often hurt each other quite seriously during combat.

YELLOW FEET AND BEAKS Steller's sea eagles have bright yellow feet and beaks.

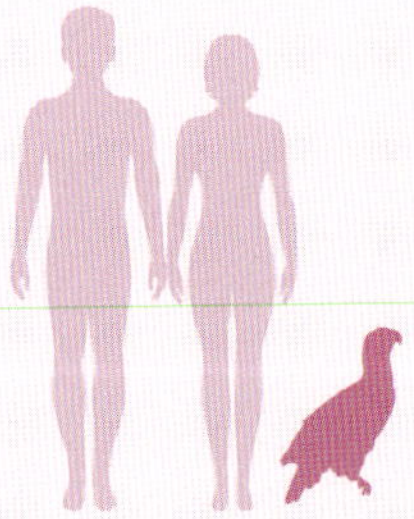

STELLER'S SEA EAGLES
(Haliaeetus pelagicus)

SIZE	85–105 cm (2.8–3.4 ft) long
WEIGHT	Up to 9 kg (20 lb)
WINGSPAN	Up to 2 m (6.5 ft)
GROUP	Convocation or constellation
HABITAT	Coastal areas and estuaries of Northeast Asia
DIET	Carnivore
LIFESPAN	Around 20–25 years in the wild
CON. STAT.	Vulnerable
KEY FACTS	Heaviest eagles in the world, but not the largest; aggressive; mainly eat salmon and trout

FISH EATERS Steller's sea eagles mainly eat fish, particularly salmon and trout, which they catch in rivers and coastal waters. They are also scavengers and opportunistic hunters, consuming carrion, seabirds and small mammals such as hares and Arctic foxes.

BREEDING GROUNDS Steller's sea eagles mainly breed along the rivers and coasts of the Kamchatka Peninsula in the far east of Russia. In winter, many migrate south to Japan and Korea, returning to the same breeding grounds in spring.

Steller's sea eagles' feet are strong and have bumpy skin called spicules. These help them hold onto slippery fish.

THE WORLD'S MOST POWERFUL RAPTOR

Harpies are the top predators in their jungle habitats where they feed mainly on sloths, opossums, monkeys and iguanas. This stealthy predator perches patiently on the lower branches of trees and waits for prey to come along. With their exceptional eyesight, they can spot a victim 200 metres (650 feet) away and fly fast through dense branches, killing instantly with the crushing grip of their talons.

HABITAT LOSS Harpy eagles need large, undisturbed areas of rainforest to hunt and nest. Their numbers have declined due to deforestation for farming and logging, as well as hunting.

CHARACTERFUL FACES Harpies have distinctive, fierce faces mainly due to a double crest of feathers that can be raised to dramatic effect. Their large, hooked beaks are perfect for tearing flesh.

CRUEL TALONS Harpy eagles' talons are 10–13 cm (4–5 in) long, or about the same size as a grizzly bear's! They use these claws to crush the brains of prey as they lift it out of the jungle and take it somewhere quiet to eat.

FAITHFUL PARTNERS Harpies mate for life, and with their long lifespans some bonded pairs stay together for 25–35 years. They breed every two to three years, working together to build a huge nest – usually in the fork of a tree – and to raise their young.

HARPY EAGLE
(*Harpia harpyja*)

SIZE	89–105 cm (31–45 in) long
WEIGHT	6–9 kg (13–20 lb)
WINGSPAN	Up to 2 m (6.5 ft)
GROUP	Convocation
HABITAT	Tropical rainforests of Central and South America, from southern Mexico to northern Argentina
DIET	Carnivore
CON. STAT.	Vulnerable
LIFESPAN	35–45 years in the wild
KEY FACTS	Huge size and strength; long talons; females larger than males

THE FASTEST ON EARTH

Peregrine falcons have moved into big cities like London, New York and Chicago, feeding on pigeons and making their homes on skyscrapers, bridges and cathedrals. They are territorial, and mated pairs often claim a skyscraper or a bridge as their own.

THE STOOP Peregrine falcons reach their top speeds of over 320 km/h (200 mph) during a stoop, or hunting dive. They fold their wings in to create an aerodynamic teardrop shape and dive steeply to strike prey in mid-air with their talons.

NO NESTS Peregrine falcons don't build nests. Instead they scratch out a shall hollow with their feet. Females lay a clutch of two to four eggs that hatch after about a month. After about two to three months the fledglings are able to hunt for themselves and become independent.

Peregrine Falcons

(*Falco peregrinus*)

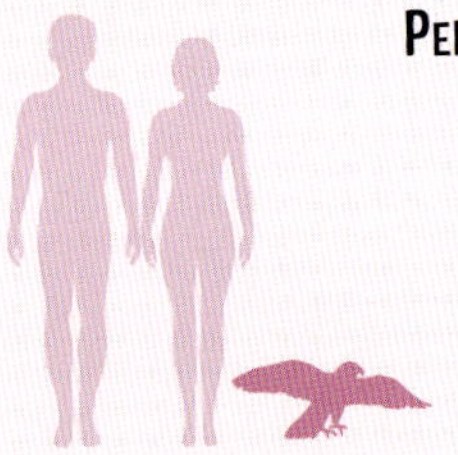

SIZE	34–58 cm (13–23 in) long
WEIGHT	Up to 1.5 kg (3.3 lb)
WINGSPAN	Up to 1.2 m (4 ft)
GROUP	Cast
HABITAT	Very adaptable, from cold tundra to hot deserts and sea level to high mountains, also urban areas
DIET	Carnivore
CON. STAT.	Least concern
LIFESPAN	6–15 years in the wild
KEY FACTS	Fastest animal in the world; captures prey mid flight

SHARP TALONS Peregrines are famous for their high-speed hunting dives but they also hunt in other ways. Often they go after their quarry in a swift horizontal chase, flapping their wings furiously. While not as fast as a nose dive, they can rival a cheetah for speed. At sea, peregrine falcons use ships, which provide high perches, to hunt for seabirds. Mated pairs often hunt together to catch their prey.

STRIKE! Peregrine falcons hit their prey mid-air with all four talons fully extended. This often produces a shower of feathers as their victims are ripped apart.

THE BIGGEST FALCONS Gyrs are the largest species of falcons. The females are about one-third bigger than the males so that they can defend their nests better.

GYRFALCON
(Falco rusticolus)

SIZE	48–65 cm (19–26 in) long
WEIGHT	0.8–2.1 kg (1.7–4.6 lb)
WINGSPAN	110–135 cm (43–53 in)
GROUP	Cast
HABITAT	Open, wind-swept landscapes, including tundra, uplands, coastal cliffs and sea ice
DIET	Carnivore
CON. STAT.	Least concern
LIFESPAN	Up to 20 years in the wild
KEY FACTS	World's largest falcon, with females much larger than males

FAITHFUL PAIRS Gyrfalcons mate for life, although they may live alone for part of the year, reuniting in the spring to breed. They are highly territorial, often returning to the same nesting site on Arctic cliffs year after year. They do not build nests, but use crevices on cliff ledges or nests that have been abandoned by other birds.

ARCTIC RAPTORS

After a successful dive at almost 320 km/h (200 mph), this gyrfalcon has its powerful, curved talons deeply buried in the feathers of its prey. These versatile birds of prey hunt by flying low and fast over the landscape to surprise prey, or by using high-speed chases and dives. They knock birds like ptarmigans, geese and gulls out of the air and onto the ground where they grab them.

EXTREME HABITAT SURVIVAL Gyrfalcons live in the coldest, most extreme habitats on Earth, from Arctic tundra and coastal cliffs to drifting sea ice in the Arctic Ocean. They have special adaptations to survive at temperatures as low as -50°C (-58°F), such as thick plumage with dense downy feathers, feathered feet, and the capacity to store fat when food is abundant. Unlike many Arctic animals, gyrfalcons do not change colour with the seasons, although they come in a range of shades from white and grey to dark blackish-brown.

UNLIKELY PREDATORS

Beautiful blue jays are large songbirds and part of the crow family. They are intelligent, noisy and complex birds that live in tight-knit social groups. While they generally eat nuts and insects, they also have a surprising 'bully bird' reputation for attacking other birds' nests and gobbling up eggs and chicks.

BLUE JAY
(*Cyanocitta cristata*)

SIZE	25–30 cm (9.8–11.8 in) long
WEIGHT	70–100 g (2.5–3.5 oz)
WINGSPAN	34–43 cm (13–17 in)
GROUP	Party or band
HABITAT	Woodlands of North America
DIET	Omnivore
CON. STAT.	Least concern
LIFESPAN	5–7 years in the wild
KEY FACTS	Good fliers; noisy birds that use a range of calls to communicate

BREEDING SEASON Blue jays are good parents. They work as a team to build the nest and raise their chicks. At breeding time they become territorial, defending the area around their nest. This is also when they attack other nests as they try to get food for their chicks.

AGGRESSIVE Blue jays can be aggressive when defending their nests and territory. They often bully other birds to control access to food. Sometimes they work together in groups to harass predators like hawks and owls; this is called mobbing.

NATURE'S GARDENERS Acorns are blue jays' favourite food. They gather them in autumn when they are abundant and bury them to eat in winter when food is scarce. But the birds forget many of the acorns they buried, so they grow into new trees.

VAMPIRE BATS

Vampire bats feed exclusively on the blood of animals such as cows, pigs, horses and birds. They hunt in the dead of night, often landing next to prey and running or jumping to attach themselves. They make a cut in the prey's skin with their special teeth and lap their blood for up to 30 minutes.

LEAF NOSES Vampire bats have special sensors on their noses that can detect where blood flows close to the skin. They have 20 razor sharp teeth that make shallow cuts in the skin of the animals they prey on. They lap up the blood, rather then sucking it out. Their saliva contains a blood thiner – called Draculin – which makes the blood flow more freely.

COLONIES Vampire bats live together in family groups called colonies, usually of around 20 to 100 individuals but often much larger. Within these close-knit communities, the bats have highly developed social lives. They groom and feed each other and often form friendships that can last for years.

SMALL BUT DEADLY Despite their small size, common vampire bats are deadly, and not just because they feed on other animals' blood. They also carry diseases such as rabies that infect farm animals and sometimes humans too. The diseases are transmitted through saliva when the bat bites their prey.

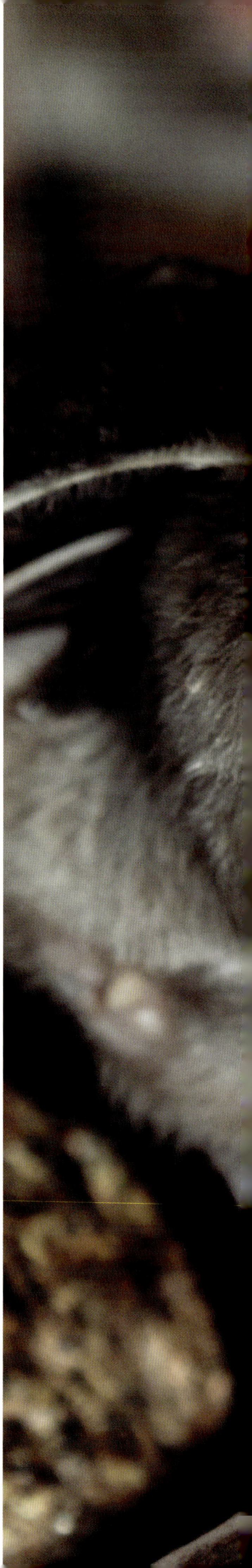

DRACULA'S BATS Common vampire bats do not normally bite humans. When they do, the bite itself is not usually harmful, but the bats can transmit harmful diseases.

COMMON VAMPIRE BAT
(*Desmodus rotundus*)

SIZE	7–9 cm (2.75–3.5 in) long
WEIGHT	15–45 g (0.5 to 1.5 oz)
WINGSPAN	15.75–20 cm (7–8 in)
GROUP	Colony
HABITAT	Tropical and subtropical regions from Mexico to Argentina
DIET	Carnivore (blood)
CON. STAT.	Least concern
LIFESPAN	Up to 12 years in the wild
KEY FACTS	Specialist predator, feeds exclusively on blood; can walk and run on all fours

ULTRASONIC HUNTERS

The brown long-eared bat hunts at night, flying slowly through the dense woodland trees and bushes and often stopping to hover. It uses its large ears to listen for prey and catches insects off leaves and bark, as well as in mid-air. When it catches a large insect, it flies to a tree branch or other perch to eat it.

ECHOLOCATION

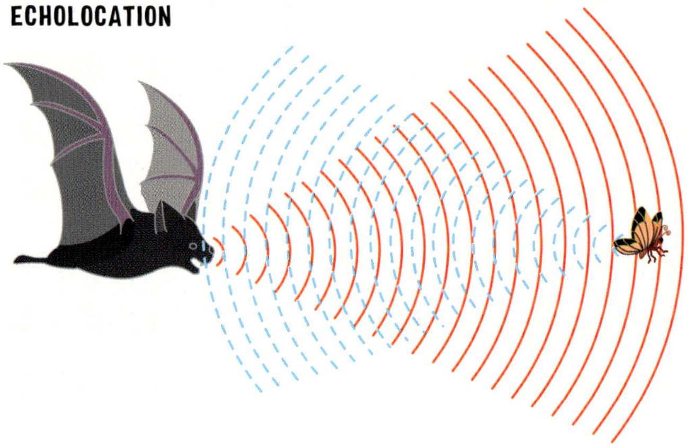

ECHOLOCATION About 70 per cent of bats use echolocation to hunt. As they fly, they emit a continuous stream of high-pitched sounds. These sounds bounce back off anything in the bat's path as echoes. By listening carefully to the echoes, the bat can work out the size and location of everything in its path, including prey.

WHISPERING BATS
Brown long-eared bats are one of a group of specialized insect-eating bats that use low-intensity, 'quiet' echolocation calls. By emitting sounds at lower decibels, they avoid alerting prey, like moths, that have evolved to hear and evade normal 'loud' bat sonar. Brown long-eared bats have large ears that act as highly sensitive, low-frequency acoustic radars for hunting.

NATURE'S PEST CONTROLLERS

Brown long-eared bats eat a lot of moths, as well as beetles, spiders, flies and earwigs. They are highly beneficial to the environment, acting as natural pest controllers by consuming large quantities of bugs, and reducing the need for chemical pesticides in gardens and on farms.

Brown Long-Eared Bat

(*Plecotus aauritus*)

SIZE	About 4–6 cm (1.57–2.36 in) long
WEIGHT	From 6–12 g (0.25–0.42 oz)
GROUP	Colony or flock
HABITAT	Deciduous woodlands, hedgerows and gardens in Europe
DIET	Carnivore
CON. STAT.	Least concern
LIFESPAN	4–5 years in the wild
KEY FACTS	Use echolocation to hunt insects; distinctive long ears; natural pest controllers

Dragonfly
(Libellula)

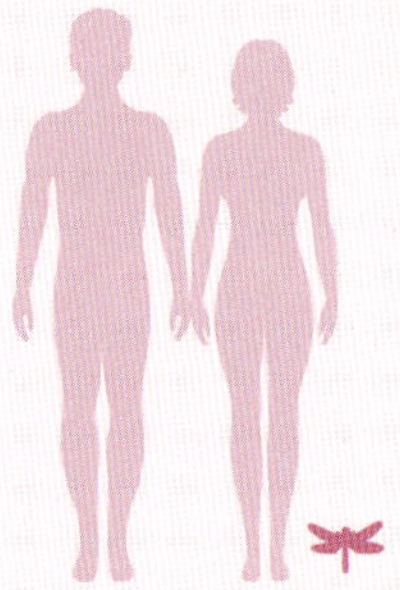

SIZE	3–8.5 cm (1.2–3.4 in) long
WINGSPAN	Up to 12 cm (4.7 in)
GROUP	Flight, cluster or swarm
HABITAT	Every continent except Antarctica, near fresh water ponds, lakes, rivers and wetlands. Often found in nearby woodlands and gardens
DIET	Carnivore
CON. STAT.	Varies, depending on species and region
LIFESPAN	1–6 months as adults
KEY FACTS	Ancient insects, they existed before the dinosaurs; expert fliers; dangerous predators

SUPERB VISION Dragonflies don't have ears so they can't hear and they have a limited sense of smell. But they do have two huge compound eyes, each one made up of about 30,000 lenses, which give them superb 360-degree vision. The compound eyes cover most of their head, as you can see in the photo above, but they also have three simple eyes, called ocelli, to help with stability and navigation.

OLDER THAN THE DINOSAURS Dragonflies are ancient creatures. They first appeared on Earth more than 300,000 million years ago, long before the dinosaurs. Some prehistoric species were very large, with wingspans of 75 cm (about 30 inches).

The beautiful dragonfly in the large photo is perched on a flower, with droplets of water in its wings.

ULTIMATE HUNTERS

Dragonflies are apex aerial predators with a hunting success rate of 97 per cent, making them one of the top predators on the planet. They are the world's fastest insects and can fly at speeds of up to 90 km/h (56 mph). Dragonflies may look beautiful to us, but they are the terror of all flying insects.

EXPERT HUNTERS When we say that dragonflies have a hunting success rate of 97 per cent, we mean that every time they begin a hunt, in 9.7 times out of ten they will successfully kill their prey. It is the highest hunting success rate of all predators.

Dragonflies usually capture prey mid-air. If it's a large victim, they sometimes carry it to a nearby perch to enjoy in peace.

AERIAL AMBUSH

Assassin flies, also known as robber flies, hunt by perching high-up on a twig so that they can scan the area around them. When they see a juicy insect approaching, they ambush it in mid-air, catching it with their legs. They inject it with a toxic mix of enzymes that make the victim's insides liquid, then they drink it, often leaving an empty carcass.

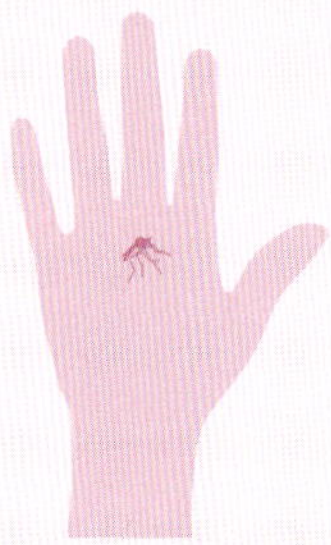

ASSASSIN FLY
(Asilidae - family name)

SIZE	5-50 mm (0.2-2 in) long
GROUP	Swarm or cloud
HABITAT	Sunny, dry, open areas like grasslands, meadows, edges of woodlands and coastal scrubs
DIET	Carnivore
CON. STAT.	Depends on the species
LIFESPAN	1-3 years, total life cycle
KEY FACTS	Over 7,000 different species; behave like the 'raptors' of the insect world

RAPTORS OF THE INSECT WORLD Assassin flies act like the raptors of the insect world, hunting from perches and catching prey in mid-air with incredible speed. Their faces are equipped with a kind of bearded armour made up of bristles that protect their eyes and head from the kicks and stings of struggling prey.

KEEN EYESIGHT AND STRENGTH Assassin flies have large compound eyes that give them almost 360-degree vision, while their long spiny legs are perfectly designed to grab struggling prey in mid-air.

INDEX

ACKNOWLEDGMENTS

The Publishers would like to thank the following picture libraries.

SHUTTERSTOCK
All photos by Shutterstock except for the following:

SCIENCE PHOTO LIBRARY
Pages 16-17, 22-23, 30-31, 33, 34-35, 40-41, 42-43